The Challenges of Space

RALPH S. COOPER
A. THEODORE FORRESTER
ARNOLD W. FRUTKIN
LEO GOLDBERG
LEONARD JAFFE
    DAVID S. JOHNSON
    WILLIAM W. KELLOGG
    JOSHUA LEDERBERG
    HOMER E. NEWELL
    AARON NOVICK

          JOEL ORLEN
          JOHN R. PIERCE
CONTRIBUTORS COLIN PITTENDRIGH
          LEONARD SCHWARTZ
          WILLIS H. SHAPLEY

    JOHN SIMPSON
    LYMAN SPITZER, JR.
    GEORGE P. SUTTON
    JAMES A. VAN ALLEN
    H. C. VAN DE HULST

GERARD DE VAUCOULEURS
ALVIN G. WAGGONER
HARRY S. WEXLER
G. P. WOOLLARD
CHRISTOPHER WRIGHT

HUGH ODISHAW
Editor

The Challenges of Space

Foreword by
EUGENE RABINOWITCH

THE UNIVERSITY OF CHICAGO PRESS

Library of Congress Catalog Card Number 62-19627

THE UNIVERSITY OF CHICAGO PRESS, CHICAGO & LONDON
The University of Toronto Press, Toronto 5, Canada
© *1962 by The University of Chicago. All rights reserved*
Published 1962, Composed and printed by THE UNIVERSITY
OF CHICAGO PRESS, *Chicago, Illinois, U.S.A.*

FOREWORD

Space exploration is an entirely novel venture of mankind, both in its reach—for the first time, man leaves the confines of his planet—and in the enormity of the effort needed to bring it "off the ground." Twenty-five years ago, on the eve of the Second World War, the discovery of nuclear fission caused the American government to sponsor a scientific enterprise then unprecedented in boldness and scope, aimed at developing as quickly as possible a nuclear weapon. In the short span of five years, the effort succeeded; and, in the process, an atomic power engine was developed as well. The spur of the war made it possible to invest $2 billion in this venture into the unknown and to mobilize on its behalf the foremost scientists of the Western coalition.

The discovery of nuclear weapons has revolutionized international relations, by imposing the need to keep peace upon nations which otherwise would have quickly and easily slid into another world war. The promise of a nuclear power for industry is fulfilling itself, but much more slowly. Nuclear submarines are being built by the dozen, but only the American merchant ship "Savannah" and the Soviet icebreaker "Lenin" demonstrate the usefulness of nuclear power for the propulsion of non-military ships. Land-based atomic power installations, initially much too costly in construction and exploitation, are becoming cheaper. In not too distant future nuclear power stations may become economically competitive with coal-based power stations. (Coal is becoming more expensive, particularly in Britain and in some other parts of the world.) If war is avoided—owing, in part, to the deterrent and sobering effect of nuclear weapons—constructive application of nuclear energy may alone, in the long run, justify the big investment made by the United States in 1940-45.

American space exploration effort, too, has been spurred by war psychology, engendered by the cold war with the Soviet Union. Big rockets, which can overcome the gravity pull of the earth and blaze the path into outer space, were first developed to carry nuclear warheads to remote corners of the world. Some even dream of using them to establish military bases on the moon and nearby planets. Without the push of military emergency, the investment in space exploration—which bids well to exceed by far the investment in the development of the nuclear bomb—probably would not have come about, at least not so soon; but if the terror of intercontinental missiles will help to hold mankind from plunging into an all-destroying war, future generations may find present investment in cosmic rocketry immensely rewarding. Bold plans to settle the moon and the planets, or at least to exploit their mineral riches, may remain science fiction; but more modest applications of cosmic rockets, such as the establishment of secure and immediate communication, by voice and picture, between any points on the globe, may alone prove profitable enough—as did the availability of tracer isotopes in the case of nuclear energy.

The main epoch-making effect of the space venture on mankind may, however, be the breakthrough from its confinement on the surface of the earth, to roaming freely through the solar system, and even beyond it—a revolutionary extension of man's realm, and a fundamental change in his psychological attitudes, rather than an increase in the material prosperity of mankind.

One of the great changes which scientific revolution has wrought in human consciousness is the blow it has dealt to the traditional earth-centered and man-centered world view, natural to the pre-scientific man. This attitude has largely survived the Copernican revolution, and the revelation by astronomy that the earth is but a tiny grain in the tiny speck of the solar system—itself a tiny speck in the galaxy of the Milky Way, which in turn, is but a minute part of the universe of innumerable galaxies. This abstract knowledge was slow in permeating the thinking of mankind, convinced of the central place of man in the divine scheme of things, and of the paramount importance of conflicts and wars between different human groups. In recent times, science has made it likely (although it has not yet proved it), that the universe contains many worlds inhabited by living organisms, including intelligent beings, with whom one day we may be able to communicate. The increasingly humble position in the universe is not easy to accept for men who have grown up in the conviction that their little human events are the chief happenings in the universe and who imagine the Creator of the world as concerned primarily with human affairs. The bodily escape, even of only a few men, from this confined world is bound to have a deep effect on the attitudes of future generations. Pride of achievement is bound to be combined, in this case, with increased humility in the face of boundless space and must stimulate a search for spiritual dignity, which alone can give man—lost as he is in the unimaginable expanse of galaxies—a feeling of worth and meaning in the universal scheme of things.

Among the most important things space exploration can do for mankind is to foster the consciousness in all men on earth of being passengers on a small vehicle, carrying them on a cosmic voyage with an unknown destination, of their community of fate and the absurdity of their quarrels.

The mastery of nuclear energy grew out of a blazing world war; the mastery of the space rocket from the blasts of the cold war; but the fierce competition between the East and the West for achievements in space exploration, now looked upon as symbols of the relative worth and capabilities of two different economic and political systems, is paving the way for a realization of the community of human interests. In contemplating the landing of men on the moon, not to speak of exploring the further reaches of the cosmos, the rivalry between groups of humanity becomes absurd. It is obviously an enterprise of mankind as a whole; its "astronomical" costs alone call for a common effort. The commitments of funds and of scientific and engineering manpower, needed for the extension of the space exploration effort to which our nations are now committed, are so high that they can put an early limitation on its scope. The Soviet Union, with its smaller productive capacity, may be the first to feel the pinch of a competitive effort in the space race, interfering with the planning for better life for its peoples and its greater industrial power; but even the United States, with their greater productivity and more ample resources, must be conscious that space exploration will compete significantly with other national enterprises, particularly in the allocation of engineering and scientific manpower. The logic of the situation, therefore, pushes all nations to the pooling of their efforts in space exploration. The need for this co-operation is obvious to scientists and to the practitioners of the space flight, the American astronauts and the Soviet cosmonauts. Gingerly but ineluctably, the governments of the two warring camps in the cold war are moving toward this co-operation. This practical education of mankind to the understanding of its essential unity is in itself worth all the investment of funds and brainpower which space exploration may require. A world war was needed for mankind to make the "quantum jump" from the world of atoms and electrons into the world of nuclei, with their million times higher energies. The cold war was needed for mankind to make the other jump, out of the sphere of earth's gravity into the weightlessness of cosmic space. Both lessons were and remain terribly costly, but learning them may bring mankind nearer to the realization of its community of interests, of its common fate. Science and technology are common enterprises of mankind, even if they are still being used (or rather misused) for the aggrandizement of this or that group. Space exploration makes the intrinsic community of scientific enterprise clear for all nations to see—and to act accordingly.

In May–June, 1961, the *Bulletin of the Atomic Scientists* published a special double issue devoted to scientific aspects of space exploration. The great interest in this issue has led to its present publication as a book. The original articles have been expanded and brought up to date, and some new ones added, so that the book is about half as large again as the original magazine issue.

EUGENE RABINOWITCH

GENERAL INTRODUCTION

The challenges of space are many, and they are perceived in various ways by different men. To a few, space activity may be no more than a spectacular pyrotechnical exhibition, unmatched by the fireworks of holidays. To others it is man's supreme adventure, dwarfing the challenge of Mount Everest and rooted in old yearnings recited in myth and history. The practical man recognizes that a large new industrial complex has quickly come into being, and he sees its applications in communications, weather forecasting, navigation, and mapping. The engineer is engrossed in the development of space systems having ever greater power and efficiency. The scientist, looking at space vehicles as carriers of measuring devices, seeks to probe far beyond the earth for new knowledge of the nature of the universe. Men concerned with political affairs see space, by virtue of its planetary and transplanetary nature, as a logical arena for co-operation among men and nations, but they also see new problems both in its military applications, and in the power of space tools to affect man's physical environment.

The space age is all of these and more. As witnesses of its beginnings, we are too close to have true perspective. If we err in our assessment of its nature and impact, it is likely to be on the conservative side. But we can take comfort in history. Who in the days of Columbus, even had he sensed the rightness of Columbus' adventure, could have predicted the Americas of today? Or who foresaw the commercial and military development of modern aeronautics in the early days of flight? Wilbur Wright in 1908 said: "I confess that in 1901 I said to my brother, Orville, that man would not fly for fifty years. Two years later we ourselves made flights." And in the same year Simon Newcomb said: "The writer cannot see how anyone who carefully weighs all that he has said can avoid the conclusion that the era when we shall take the flyer as we now take the train belongs to dreamland."

This book is not concerned with prophecy. It is largely an exposition of what exists today and what is technically and scientifically within reach. Its purpose is twofold: to outline the variety of activities—technical and scientific, domestic and international—that make up man's space endeavors, and to stimulate analysis of these activities. The writers are specialists in their fields, and their objectives are basically expository. The sum of their contributions is designed to give a reasonably well-rounded description of the technological, applied, and research aspects of space efforts; to summarize the national space programs throughout the world; and to outline the mechanisms and activities in international co-operation.

One of the remarkable aspects of the space age is the sheer level of accomplishment attained in a few years: more than one hundred artificial earth satellites and deep-space probes have been launched in five years. Of these, twenty were launched by the Soviet Union and more than eighty by the United States. The Russians have been sensitive to the importance of achieving historical feats in space. The first artificial earth satellite was Sputnik I, launched on October 4, 1957. The first photographs of the back side of the moon were made by Lunik III. The first manned orbital flights were Russian. The Soviet effort has emphasized man in space, and, of their first twenty launchings, at least twelve have been related to this program. Research has assumed a secondary role; most of it was done on the first few Sputniks and Luniks.

Of the eighty or so United States space ventures during the past five years, half were launchings for the military program, half for the civilian program. Most of the former have been Air Force efforts, largely in the Discoverer series; these launchings have been concerned with engineering problems, with the recovery of capsules, and with the development of reconnaissance and early warning systems, although some scientific work has also been done when weight requirements permitted capsules to carry more instruments.

The civilian effort has been the responsibility of the National Aeronautics and Space Administration (NASA). The Explorer, Pioneer, Ranger, Tiros, and Mercury launchings have been mainly part of the NASA program, which has ranged over the principal fields of space activity. There have been several sub-orbital and orbital flights as part of NASA's man-in-space endeavor. There have been significant accomplishments in satellite applications for geodetic, meterological, and communications purposes. The first cloud cover satellite, Vanguard II, was followed by several successful orbitings of the Tiros series, and data from these satellites have been used for storm warnings and weather forecasting. The first active and passive satellite communications tests were performed by the United States. The scientific satellites and space probes have been particularly extensive and rich in results: the discovery of the Van Allen radiation belts; the development of the "pear-shaped" model of the earth, which is based on perterbations in the orbit of Vanguard I; and the correlation between drag on a satellite and solar flux are typical of the early findings.

The paragraphs above touch upon the major themes of space activity, and it is to the development of these themes that this book is addressed. More explicity now, we can identify three major aspects of space enterprise (for science is not the only claimant upon space, even though it is invoked on every occasion). These aspects are adventure and exploration, satellite applications, and research.

True enough there are linkages among these areas. All depend, to varying degrees, upon space systems. Applied research, development, and engineering that deal with past, present, and future vehicle systems for one area of activity cannot be divorced from other areas. This is also true of the tracking and telemetry network of stations. Beyond technological and operational interrelationships, moreover, there are other linkages. Meteorological satellites provide storm warning and cloud cover data useful to the weather forecaster, but these data provide the research meteorologist with information on the dynamics of the atmosphere; geodetic satellites are designed for navigational and mapping objectives, but orbital analyses shed light on the shape and structure of the earth; communications satellites will perform service functions, but the ionospheric physicists study the propagation of radio waves used in these ways to elucidate general questions about the waves, the ionosphere, and the atmosphere.

The man-in-space program for some years promises little if anything in adding to our knowledge of the universe, but ultimately the study of the moon and nearer planets will depend on manned expeditions. Pure research ostensibly contributes little in the short run to application or adventure, but all that can be learned from instrumented devices about the physical environment in space and around or at the surface of the other bodies in the solar system contributes to the intelligent design of missions, manned or unmanned, to the moon, Mars, or Venus. Obviously one factor in these interrelationships is time: a manned expedition to Mars, carrying suitable instruments, will be a great scientific expedition as well as a grand human adventure. At present, however, each of the three principal areas of space interest has its own rationale, established by its purpose.

The man-in-space program may, for our purposes, be considered a program of gross exploration—"gross" only to distinguish its early phases from those that follow, which will include quantitative scientific investigations. In many ways we are in a pre-exploration period, for man can now do little actual exploring, even in the sense of geographical explorers of old, bent on sheer discovery or riches. The program is concerned with getting man into space, with developing more ambitious manned spacecraft, with trying out and developing man's capabilities in a novel and dangerous situation. The purpose of the program—a purpose which will probably exist for a decade—is the development of human capability in space. However much we may link this capability in imagination or hope to its eventual exploitation in exploration and research, the pursuit of the capability is presently tied to public interest in human exploits and to political interests in world leadership. In many ways national interests in space exploration parallel those that led nations to compete in geographical discovery.

The field of satellite applications has the simplest of all purposes: the development of communication, forecasting, navigation, and surveying systems of immediate, practical use. No new principles of science are called for in attaining these systems, and thus they lie within reach, though the engineering, developmental, design, and operational problems are neither simple nor few.

GENERAL INTRODUCTION xiii

The claims of science hinge upon man's passion for knowledge about the cosmos, of which he and his planet are integral parts. The results of the balloon, sounding rocket, satellite, and space probe research efforts of the last decade reveal that the vehicles of the space age are powerful tools to this end. In the past, man has been tied to earth, and all that he has learned of the universe has depended on observations of his planet, on ground-based observations of all that lies beyond earth, and on the power of his mind and imagination in analysis and interpretation. Now he has the ability to reach beyond his old limits, to sense and measure directly matter, events, and processes in space.

To these three areas of space activity, we must add a fourth—space technology itself. Obviously space activity depends upon space systems. The Soviet Union, by virtue of an earlier start, was not only first into space with Sputnik I but had engines capable of launching heavy payloads. The heaviest payload placed into orbit by the Soviet Union was Sputnik VII, launched on February 4, 1961; it weighed some seven tons. The largest United States satellite was launched on February 20, 1962, and carried John Glenn on the first United States manned orbital flight; it weighed about a ton and a half. The development of more powerful rocket systems for space is an important part of the United States program, for out of it will come the more powerful tools needed for both manned and unmanned endeavors of the future.

Space technology is significant not just because it is the means to certain ends but also because it is a very large human activity. The development of rocket systems for space launchings is a multibillion dollar affair. It includes the design and construction of rocket engines, both solid and liquid; the fuels, their containers, pumping and control equipment; the auxiliary devices associated with guidance, tracking, monitoring, orientation, and stabilization. It includes extensive launching and tracking facilities. It has generated a vast new industry, begun recently and still growing. It is, in short, a major factor on the industrial and economic scene.

To each of these four topics—exploration, application, research, and technology—a section of several chapters is devoted with one exception. Exploration by man is dealt with, explicitly and implicity, in several sections, because it is still in its developmental phase and because its accomplishments, either in terms of the courage of individuals or the capability of technology, are widely known. In addition, two separate sections of the book outline national programs in space and international aspects of co-operation in space. The chapters on national programs make clear the scope of our own space effort, summarize that of the Soviet Union, and suggest the significance of the growing space activities in nations that do not yet have launching capabilities. The chapters on international aspects of space work are intended to be informative. There is a growing interest in space on the part of intergovernmental bodies focused within the United Nations. The co-operative research of the International Geophysical Year is a continuing activity within the scientific community.

In each of these areas much has happened but much more remains to be done. A technology capable of taking man to the moon and back does not yet exist. Fully operational systems in, say, communications are still being developed. While the results of space research have been rich in quantity and quality, the field has barely been touched. A good start in space co-operation was one of the legacies of the International Geophysical Year. Governments and intergovernmental organizations are showing increased interest in the public and international problems of space activity. It is to these outstanding questions and problems—the several challenges of space—that this volume is addressed.

ACKNOWLEDGMENTS

Above all, acknowledgment is due the authors who set aside their own pressing work to turn to the writing of these essays in fields of their special competence. To Dr. Rabinowitch, the editor of the *Bulletin of The Atomic Scientists,* I owe the original assignment of compiling the special issue of May–June, 1961. Mrs. Ruth Adams, Associate Editor of the *Bulletin,* helped with wise counsel and hard work in both the *Bulletin* enterprise and in this book version. Miss Paula Fozzy, Assistant Editor of the *Bulletin,* assumed many editorial burdens for the book version. My secretary, Miss Grace C. Marshall, spent many evenings and weekends typing, checking, reading manuscript and proof and compiling the indexes. To all of these, and to others whose names I have not mentioned, I express grateful thanks.

HUGH ODISHAW

CONTENTS

PART ONE:
APPLICATIONS OF SPACE RESEARCH
- 3   Introduction   Hugh Odishaw
- 7   Meteorological Satellites   Harry Wexler and David S. Johnson
- 24   Space Research and the Earth Sciences   G. P. Woollard
- 44   Communications Satellite Systems   Leonard Jaffe
- 60   Hazards of Communications Satellites   John R. Pierce

PART TWO:
SPACE RESEARCH
- 75   Introduction   John A. Simpson
- 81   Biology and the Space Environment   Colin S. Pittendrigh
- 89   Challenges to Biology   Aaron Novick and Joshua Lederberg
- 97   Flying Telescopes   Lyman Spitzer, Jr.
- 108   Rocket Probes   William W. Kellogg
- 118   The Earth and Near Space   James A. Van Allen
- 129   The Sun   Leo Goldberg
- 142   The Moon and Planets   Gerard de Vaucouleurs

PART THREE:
NATIONAL SPACE PROGRAMS
- 155   Introduction   Hugh Odishaw
- 161   United States Space Program   Willis H. Shapley
- 178   NASA and Space   Homer E. Newell
- 195   Department of Defense Space Program   A. G. Waggoner
- 204   Space Programs of Other Nations   Joel Orlen

PART FOUR:
INTERNATIONAL SPACE CO-OPERATION
- 235 Introduction — Hugh Odishaw
- 241 International Space Organizations — Leonard E. Schwartz
- 267 International Programs of NASA — Arnold W. Frutkin
- 277 The United Nations and Outer Space — Christopher Wright
- 291 COSPAR and Space Co-operation — H. C. van de Hulst

PART FIVE:
SPACE TECHNOLOGY
- 301 Introduction — Hugh Odishaw
- 304 Space Vehicles — George P. Sutton
- 319 Chemical and Nuclear Rocket Propulson — Ralph S. Cooper
- 336 Deep Space Propulsion Systems — A. Theodore Forrester

- 355 AUTHORS

- 363 INDEX OF NAMES

- 365 SUBJECT INDEX

PART ONE

Applications of Space Research

Introduction
Meteorological Satellites
Space Research and the Earth Sciences
Communications Satellite Systems
Hazards of Communications Satellites

INTRODUCTION

HUGH ODISHAW

Three major current uses for artificial earth satellites are communications, meteorology, and geodesy. In each of these areas the application is within reach because the scientific principles are old, even though the technology is new. In each area, too, some experience had been obtained.

In communications, the possible uses of satellites embrace radio telephony, telegraphy, and programmatic broadcasting, both of radio and television. The immediate reason for pursuing satellite transmission is the limited capacities of present message services. The impetus is largely two fold. First, the presently available range of frequencies for long-range radio communications is restricted to those that are reflected by the ionosphere, for it is by reflection, whether single or multiple, that radio messages can be sent over long paths. But the usable frequency spectrum is overcrowded. Room for expansion exists in the shorter-wave region; here the waves pass through the ionosphere and cannot be used for long-range purposes by traditional techniques, but they are suited to satellite methods. Second, growing needs, particularly in telephony, make attractive the possibilities of communications satellites, both nationally and internationally. For example, while the population of the United States has increased about 60 per cent and the number of telephones 500 per cent during the last forty years, the number of telephones abroad has risen by almost 800 per cent with a world population increase similar to that of the United States. Moreover, while domestic telephone calls rose from 17 billion to almost 88 billion during that period, long-distance messages handled by the American Telephone and Telegraph Company increased from 16 to 532 million, an increase of more than 3,000 per cent.

All satellites and space probes have carried specialized radio receiving and transmitting equipment, in part for guidance and control purposes, in part to relay experimental data from space to earth. The first satellite transmission of a human voice was achieved by the United States on a satellite launched by an Atlas rocket December 18, 1958. A recorded message carried aboard the satellite was transmitted, but messages were also accepted and relayed from ground stations in Texas, Arizona, and Georgia. The development of practical operational systems, after this historic and experimental event, then got under way.

In meteorology, the satellite application of immediate interest is weather forecasting, including monitoring major storms. The promise of weather-forecasting satellites has already been realized, and may be significantly valuable to mankind. Such systems can detect and track storms, almost from their birth to their decay, and provide information that permits man to take protective measures. The technique consists essentially of photographing clouds all over the earth; the television pictures are transmitted to ground stations; analyses of cloud cover provide information on atmospheric movements and disturbances. The first satellite to attempt cloud-cover photography was Vanguard II, which, during the IGY program, used photocells to produce crude images. Since then, Tiros I (launched March 1, 1960), Tiros II (launched November 23, 1960), and Tiros III (launched February 8, 1962) have provided extensive photographic coverage of cloud cover. Advanced systems, in the forthcoming Nimbus series, are being developed.

Although these data are immediately useful in surveying cloud cover, and in tracking major disturbances and patterns, they are also valuable for research because they yield information on atmospheric circulation and dynamics. Such observations, when coupled with other information from ground-based and satellite-borne experiments, afford hope that an adequate understanding of meteorology may be attained in the next few decades. Better understanding of weather phenomena is needed for any future attempts by man himself to affect weather.

The science of geodesy is the third major area of satellite applications. Geodetic satellites are of interest both for practical purposes and for research. Navigation and surveying are the

Introduction

principal practical objectives, for geodetic satellites can become standard, universal devices for surface and air navigation, and they could be used also for surveying the surface of the earth, thus tying together geodetic survey networks throughout the world. At the same time, geodetic satellites serve research in studies of gravity, the shape of the earth, and even its composition and structure. Thus the first small Vanguard satellite (Vanguard I, launched on March 17, 1958; a trifle more than six inches in diameter and three pounds in weight; still in orbit) provided much orbital data. Analysis of its paths about the earth permitted American scientists to postulate a new, slightly "pear-shaped" form for the earth. The variation from the previous ellipsoidal theoretical model is extremely small, but the new model provides better notions of the earth's structure and mass distribution.

In addition to the preceding three applications of space science there are others. One which is now being developed actively is the field of military application. Of course, military services will make use of communications, weather, and navigational and surveying satellites; in so doing they are only another group of users. But there are also uniquely military prospects. Two have been under development: reconnaissance and early warning satellite systems. Moreover, in principle, space systems could carry explosive devices which could be ejected from the space carrier and directed to earth targets. The development of such weapon systems would be costly, and the same end could be far more readily achieved by fixed or mobile surface installations. But the fact that no scientific principle bars their development, coupled with that political suspicion which leads nations to pursue a course of action if another nation might also be pursuing it, suggests that mankind may be faced with military space systems. For history shows that mankind has not refrained from exploiting technical potentialities, as they became apparent, from the spear to bomb-armed guided missiles.

The facts that space carriers can travel about the earth at great distances—and even, at some 24,000 miles away, remain poised above a point, by traveling at the same velocity as the earth—and that they can eject matter into the high atmosphere and near space, raise possibilities of other applications. These may be roughly classed as applications that affect earth's spatial environment. For example, man has injected into the upper atmosphere with sounding rockets more sodium than existed there naturally. The Argus experiment, in which small atomic devices were carried high above the earth by rockets and detonated there, injected charged particles which affected the earth's magnetic field temporarily, interfered with communications, and created artificial auroras. The Russians have suggested creating a light-reflecting layer far enough from the earth so that some sunlight would be reflected, lighting the dark side. Discussions of weather control raise the question of how satellites might eject matter affecting, say, cloud cover and storms. Such prospects, some realized, some remote, some perhaps unrealizable, pose critical problems.

METEOROLOGICAL SATELLITES

HARRY WEXLER and DAVID S. JOHNSON

For the first time in history, there exists an observing platform which can detect atmospheric conditions long before local meteorologists relying on conventional techniques may be aware of them. This platform is the meteorological satellite, which, even in its present primitive stage, has already contributed significantly to meteorological developments by depicting cloud systems and interpreting them for daily weather prediction, and by collecting basic physical data such as measurements of the radiative exchange between the earth and sun and space. Future observations will include the temperatures of cloud tops and the earth's surface, the average temperatures of layers of the clear atmosphere, concentrations of water vapor, ozone, and other properties not yet envisaged.

Man is immersed in a working fluid of a global experiment—the earth's atmosphere—a fluid so massive that there are nearly two million tons of it for each person on earth. From above, it is penetrated by energetic particles and radiations, and from beneath, deformed, restrained, heated, and cooled as it passes over the irregular earth surface in its endless quest to equalize its energy imbalances, thus creating wind and weather. The atmosphere performs countless cycles of interrelated phenomena of every size, from global to microscopic. They are all important. For example, those actions involving water vapor—which comprises only about 0.2 per cent of the total mass of the atmosphere—nevertheless have such a profound effect on our planet's heat balance that without them the mean temperature of the earth would drop by 40 degrees centigrade.

Meteorologists have traditionally been handicapped by having only fragmentary knowledge of what is going on in the atmosphere at any time. About a century ago, national meteorological services were established to provide forecasts to the public. As observing networks expanded geographically and in altitude, meteorologists continued their audacious attempts to predict the future state of a three-dimensional system whose initial state was inadequately known. Because of insistent public demand, the forecaster makes his daily predictions and up to a certain point is generally successful. His successes, however, are generally limited to forecasts for not more than a few days in the future, and for areas in the midst of, or close to, a fairly dense observing network so that unknown disturbances from distant and sparsely observed regions have not had time to exert significant influence. Even so, disturbances such as tornadoes or severe thunderstorms can develop suddenly or slip through the mesh of observing stations.

Pictures of such storms, continuous in space and time, can now be provided by the proper distribution of surface radars and meteorological satellites. Radar observations show the three-dimensional array of many cloud types, particularly those with precipitating particles, over limited areas (200–300-kilometer radius) surrounding the radar set. In earth-orbiting satellites, meteorologists have for the first time an observing platform with a capability commensurate with the global extent of the atmosphere, not only for observing what is going on but for communicating observations and warnings over the entire world.

Until recently, high altitude photographs of large cloud systems for meterological research have been available as the result of relatively few rocket flights. The V-2 rocket photographs at White Sands, New Mexico, in 1947 were among the first; then followed the Aerobee photograph of an upper cyclonic vortex over Texas in October, 1954. In August, 1959, pictures obtained from an Atlas nose cone showed a frontal cloud system 3,000 kilometers long, extensive lines of tropical clouds, and cellular cloud patterns over the north Atlantic ocean. In April–June, 1960, 14,000 high-quality cloud photographs were obtained over the world between 55 degrees north and 55 degrees south latitude by the meteorological satellite Tiros I, launched by the National Aeronautics and Space Administration. These revealed spiral cloud bands associated with cyclonic vortices 800 to 1,500 kilometers in diameter, cellular arrangements of cumulo-

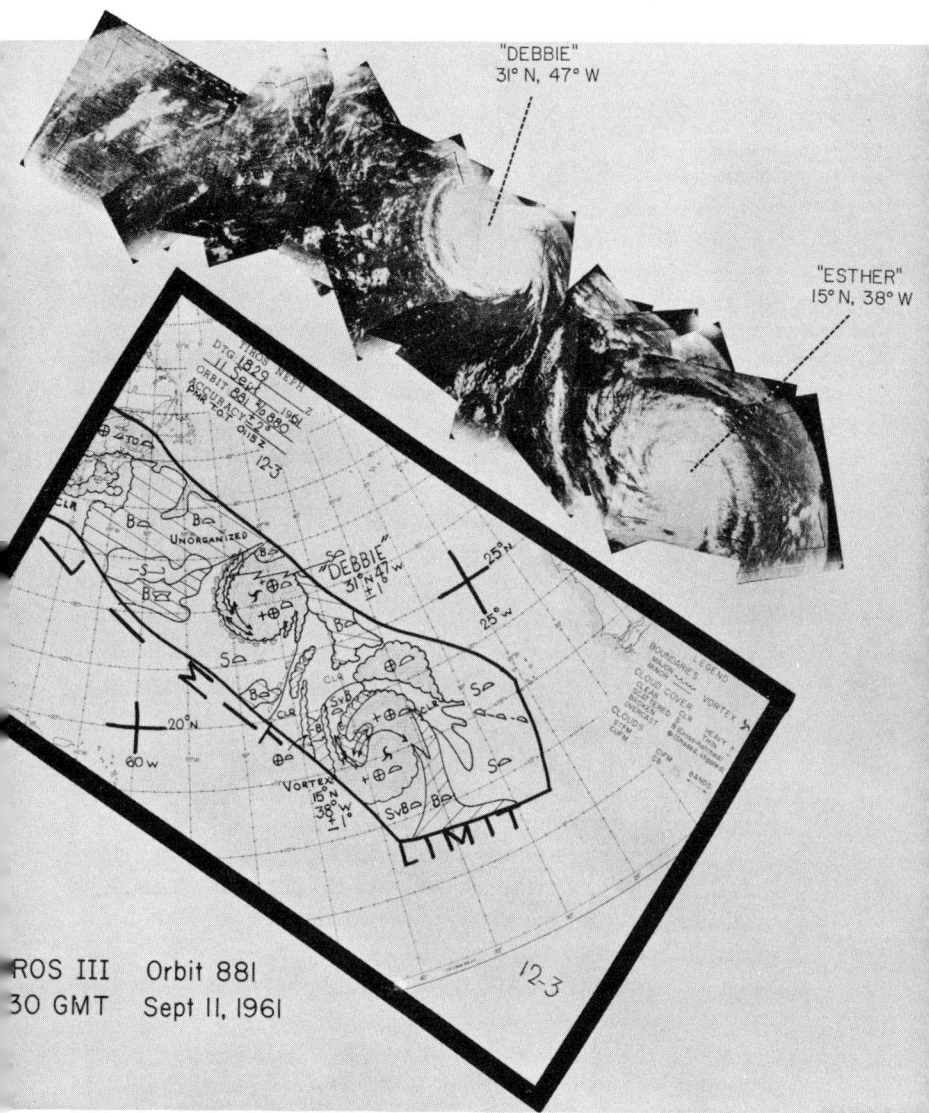

Fig. 1—The cloud pictures obtained by Tiros III during part of one orbit on September 11, 1961 are shown in the mosaic in the top portion of the figure. The hurricanes "Debbie" and "Esther" are clearly shown. The cloud analysis based on these pictures (shown at the bottom) was prepared by meteorologists at the data-receiving station within a few hours after receiving the pictures from the satellite. These cloud maps are transmitted by weather facsimile circuits to forecast centers. (Mosaic courtesy of C. Erickson.)

form clouds 50 to 80 kilometers in diameter resembling in appearance the Bénard cells studied in the laboratory, and frontal cloud patterns which confirm in a remarkable manner the classical Norwegian model of cyclone families. In general, the cloud pictures from the Tiros satellites reveal an unexpectedly high degree of organization, and unsuspected scales of local activity in atmospheric motions. Full interpretation of these natural "weather maps" will require intensified empirical, laboratory, and theoretical research in the dynamics of convection and vortex flow.

The Tiros I photographs also illustrated in a dramatic way the potential usefulness of satellites in the detection of severe storms by photographing the unique spiral cloud structure of a hurricane near New Zealand, and a large cumulonimbus cloud system in the midwestern United States which later produced tornadoes. Snow and ice areas over land and oceans were also observed.

Cloud observations from satellites, once they serve their immediate forecast use, are valuable in meteorological research. They can be used to establish for the first time a truly global cloud census, to draw average charts of world cloud cover by months or other periods, and to note long time variations in cloud amount and distribution. The world cloud cover is the most important component in determining the earth's albedo, and serves as a natural thermostat to keep the world temperature within narrow limits. A larger than average cloud cover can reflect more solar radiation and cool the earth, thus reducing the convection currents and cloud cover. A smaller cloud cover will enable more solar radiation to heat the earth's surface and thus cause more clouds to form.

A second Tiros satellite was successfully placed in orbit on November 23, 1960, and produced useful cloud photographs for more than six months. Tiros III was launched on July 12, 1961, and was still providing excellent pictures after nearly three months of operation. The National Aeronautics and Space Administration plans to launch four more Tiros satellites before the end of 1962.

The operation of Tiros III during the hurricane season has provided spectacular pictures of hurricanes and other tropical

Fig. 2—Hurricane "Betsy" was photographed by Tiros III on September 8, 1961, when the storm was centered at 36° north latitude and 59° west longitude. This picture illustrates dramatically the cyclonic circulation associated with the hurricane.

storms. For example, hurricane Esther was photographed by the satellite two days before its existence was confirmed by conventional observations. These satellite pictures are being studied to trace the evolution of tropical storms as manifested by their cloud patterns.

In addition to the television cameras identical to those used on Tiros I, Tiros II carries two sets of radiation sensors. One radiation instrument relies on the motion of the satellite to scan the earth with five sensors having a spatial resolution of about 50 kilometers and designed to measure reflected solar radiation from the earth and hence the albedo; infrared radiation from the earth; emission of the atmosphere in the 6.3-micron water vapor band;

temperature of the earth's surface or the cloud tops; low resolution cloud cover in visible light for comparison.

In the second experiment, two sensors, having a resolution of about 800 kilometers, view a portion of the earth in the center of the area depicted by the wide-angle television camera. One sensor responds to the terrestrial radiation emitted by the earth and atmosphere, while the other measures this radiation as well as the reflected and scattered solar radiation.

An additional set of radiation sensors was carried on Tiros III, which also responded to the terrestrial radiation and the terrestrial plus reflected and scattered solar radiation. These sensors are omnidirectional; their field of view is that part of the earth bounded by the horizon as seen from the satellite.

Observations from these radiation sensors are still under study, but a preliminary analysis indicates that much of the data are of excellent quality and should reveal useful information concerning the radiative sources and sinks in the atmosphere, the surface and cloud temperatures and thus the approximate heights of cloud tops. The radiation sensors on Tiros II and III represent a continuation and extension of the low resolution solar and terrestrial radiation measurements made from the Explorer VII satellite, which was launched on October 13, 1959. The data for Explorer VII are still in process of analysis, but some interesting preliminary charts of outgoing radiation from the earth to space have already been drawn.

These satellite observations of radiation will be used to study weather developments and motions of atmospheric disturbances as related to the radiation balance, and to compute net radiation gains and losses over the earth—a possible aid to long-range weather prediction. It is known that poleward transport of excess energy from the tropics can vary considerably. When winds are blowing mostly from the west or east, the energy flow away from the tropics is inhibited. But when the flow pattern changes from this "high index" condition to meridional or "low index" flow pattern (typically characterized by large quasi-stationary anticyclonic and cyclonic vortices), poleward energy flow is promoted, giving rise to rather prolonged spells of the same general weather type — fair weather, droughts, floods, storms — depending on geographic location with respect to the stalled weather pattern.

Observations of unusual solar radiations, energetic particles, and meteoric dust from space should be readily available from satellites for correlation with unusual weather behavior. World weather charts should make it possible to study contrasts and similarities in northern and southern hemispheric circulation patterns, and to see whether one hemisphere precedes the other in assuming anomalous patterns or whether both react simultaneously to a common external excitation such as unusual solar radiations or meteoric dust. Also, observation of compositions and circulations of atmospheres of other planets by space probes, planetary satellites, and landings, will permit comparisons with the terrestrial atmosphere and will assist in separating phenomena common to all planetary atmospheres from those characteristic of a particular planet.

CLIMATIC CHANGES

Much climatic warming, which has attracted considerable attention in recent years, is confined to rather restricted areas of the earth, particularly to the north Atlantic subarctic region and the eastern and central United States. Other regions, such as the northwestern United States and parts of central Canada, have cooled appreciably during the past forty or fifty years. But whether the earth as a whole is warming or cooling cannot be determined from the present inadequate observation networks. Measurements over periods of a year or longer of the incoming solar radiation and the fraction that is reflected by the atmosphere, clouds, and earth's surface (earth albedo), as well as the outgoing infrared radiation from the earth and the atmosphere, should enable us to learn whether the earth is receiving more energy than it returns to space or vice versa.

The difficulties of detecting changes in energy budget of some of these components by conventional means are illustrated by the following. On the average, each square centimeter of the earth's surface and the atmospheric column above it annually receives from the sun about 175,000 gram-calories. If there were complete radiative balance (as most textbooks in meterology assume) there should be an equal return of energy to space. But if, averaged over a year, there were a 1 per cent excess of energy received by the earth, what would this mean in terms of heating the oceans, melting the glaciers, and warming the atmosphere? If all the extra energy were concentrated in heating the oceans and nothing else, this would raise their average temperature by 0.006 degrees centigrade, which could not be detected by present observing techniques.

What if the excess energy resulting from the imbalance of incoming and outgoing energy streams were concentrated in the atmosphere? This would raise the average atmospheric temperature by 7 degrees centigrade—an amount believed to be the difference in world air temperature between that of the last ice age and the present time—a difference readily detectable even by our present observation networks.

Actually the excess (or deficit) of energy would be partitioned among atmosphere, oceans, glaciers, and land in some unknown manner. Determination of the over-all global energy budget plus that of such individual components as the atmosphere and glaciers (with the aid of greatly expanded observation networks) would permit determination of the energy budget changes of oceans and land masses. Thus the advent of satellites will unveil a new tool by which meteorologists may soon learn more about the energy exchanges of the planet earth with sun and space, and how the exchanges, in turn, may be translated into the climate trends on which man's existence depends so crucially.

OPERATIONAL USE OF SATELLITE DATA

Although significant improvements in meteorological analysis and forecasting will most likely result from a broad research program to which meteorological satellite data will contribute, it has already been possible to utilize cloud data from the Tiros satellites to a limited degree in meteorological operations. While operational use of data is considered to be experimental, it has already indicated the potential of even such simple analyses as

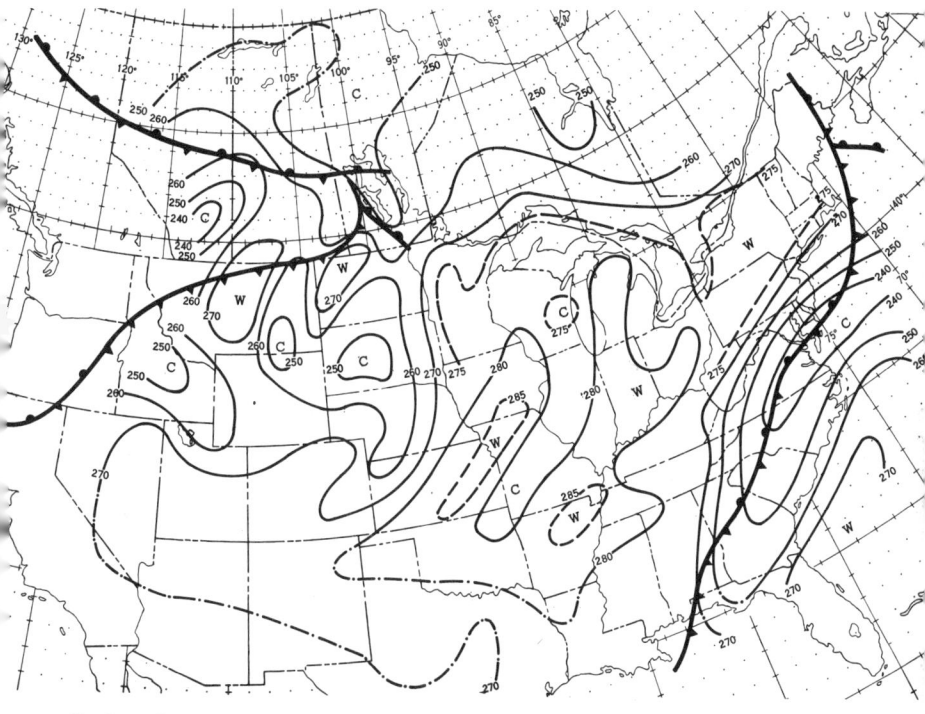

FIG. 3—A map based on the radiation emitted in the 8-to-12-micron-water-vapor "window" by the earth's surface and the top of clouds as measured by Tiros II on November 23, 1960. The contours are in degrees Kelvin. The weather fronts are shown for comparison. In overcast areas where the cloud tops are high, such as along the east coast, low radiation values are obtained. The highest radiation temperatures are measured in clear areas such as in the Midwest in this example. (Courtesy of S. Fritz and J. S. Winston.)

cloud maps. They are particularly valuable in filling in the vast distances on weather maps between existing conventional observing stations, particularly over the oceans and other data-sparse areas of the world.

While the radiation data from the Tiros satellites are not yet being used in meteorological operations, preliminary study of these data indicates they will be operationally useful as soon as rapid processing and analysis techniques are established. For example, the radiation measurements made in the 8–12-micron water vapor "window" show promise of indicating in a gross sense the world's cloud distribution at night, complementing the existing daytime cloud observations obtained by the vidicon type television cameras. The correlation between the various radiation temperatures and weather patterns indicates that these data will be operationally useful in an empirical sense even though much research is required before the full import of these data in a quantitative, physical sense will be known.

The nature of the orbit and type of stabilization of the Tiros satellites do not permit the observation of the whole globe on a daily basis. The Nimbus satellite now under development by the National Aeronautics and Space Administration is designed to overcome the limitations in observational coverage inherent in the Tiros satellite. The quasi-polar orbit planned for Nimbus will bring the entire earth into its view twice each day—a great impetus to the operational use of satellite data. By an active stabilization system, one axis of the satellite will always be normal to the earth's surface.

METEOROLOGICAL SATELLITE SYSTEM

The initial operational system will utilize a single Nimbus-type satellite launched so that on each orbit the satellite will cross the equator at the ascending node (northbound) at local noon and again at the descending node at local midnight. Television cameras will produce cloud pictures over the entire solar illuminated portion of the earth once each day with a resolution of about 0.8 to 2.4 kilometers. The cloud distribution over the portion of earth in darkness can be determined by a scanning infrared radiometer.

As an extension of the operational system, a second Nimbus satellite might be launched so that one satellite crosses the equator at the ascending node at 9 A.M. local solar time and the second at 3 P.M. Thus each point on the earth's surface would be viewed at least once every six hours, twice in daylight (except for the winter polar regions) and twice in darkness (except for the summer polar regions).

The complexity and tremendous volume of data expected

Meteorological Satellites

from even the early experimental Nimbus satellites will require the use of a costly ground station for satellite control and data reception. Only one such station would be required if it were located at a point poleward of about 80 degrees latitude. Initially, two or more ground stations probably will be used, utilizing existing facilities as much as possible. Wide bandwidth communications will be required to speed all the data to a central analysis facility, such as the National Meteorological Center in Washington, D.C. Meteorologists, assisted by high-speed data processing equipment, will reduce the observational data and prepare analyses for both national and international use. The analyzed satellite observations will be distributed to field forecast centers by teletypewriter and high quality facsimile equipment.

While limited amounts of satellite data can be transmitted to other nations of the world via existing teletypewriter circuits, the global nature of satellite observations indicates the need for more adequate global meteorological communications. Indeed, the large volume of data expected from meteorological satellites, when added to the ever increasing number of conventional meteorological observations, sharply emphasizes the urgency of providing greatly improved communications in support of meteorology. Communication satellites now under development offer promise of providing a means for the world-wide dissemination of meteorological data for the benefit of all mankind.

It also appears feasible to measure the vertical distribution of precipitation beneath the satellite by using radar carried by polar orbiting satellites. Such data would greatly enhance the value of the cloud pictures because of the added knowledge of vertical distribution of cloud layers not obtainable from the cloud pictures alone, and the ability to determine something of the intensity of storm systems as indicated by the intensity of precipitation. The height of the freezing level in the atmosphere could also be inferred from the characteristic radar echo produced when ice and snow begin to melt after falling below the freezing level in the atmosphere. Further engineering development may ultimately permit the use of radar for detection of clouds, including measurements of the height of the bases and tops of various cloud layers beneath the satellite.

Theoretical studies have recently shown that the mean temperature of several layers of the atmosphere in at least the high troposphere and stratosphere can be determined from a series of high resolution spectrometric measurements of the radiation emitted by the earth's atmosphere in various portions of the 15-micron carbon dioxide band. A spectrometer for this experiment is now under development. The same spectrometer is also being designed to measure the radiation received in a narrow spectral interval centered at 11.1 microns. The gaseous atmosphere is transparent to radiation at this wave length, so that the radiation received by the instrument would be proportional to the temperature of the earth's surface or cloud tops within the field of view of the instrument.

Great strides have been made in recent years in utilizing digital computers to predict the weather by solving numerically simplified forms of the equations of atmospheric motion. The procedures presently used require quantitative observations as input to the computer. The aforementioned temperature measurements will probably contribute to numerical prediction, as will the vertical velocity field as deduced from cloud pictures.

Present numerical prediction models require upper air wind velocities and some parameter directly related to the height of selected constant pressure surfaces in the atmosphere. It has been suggested that balloons be injected into the atmosphere which will move with the wind on surfaces of constant density. A network of balloons circling the globe at three different levels and equipped with temperature sensors would go a long way toward permitting numerical prediction on a global basis. Polar-orbiting meteorological satellites might be used for locating the balloons with respect to the earth's surface and receiving the temperature data from the balloons. The change in the position of a given balloon between observations would yield the mean wind during that period. The observational data could be processed at a central location before international dissemination.

A polar-orbiting satellite system of the Nimbus type, while providing rather frequent observations of the whole globe, does not meet the meteorologist's objective of continuously monitoring the world's weather. A satellite in an equatorial orbit and synchronous rotation with the earth, 35,700 kilometers above the surface, will appear to be stationary. Four such satellites equally spaced about the equator would provide a system of observing platforms capable of viewing all of the earth's surface continuously in the zone between 60 degrees north and south latitude. The observational coverage of these "stationary" satellites would be complemented in the polar areas by the Nimbus type of satellite.

It appears feasible to obtain cloud observations only from stationary satellites because of their great altitude above the earth's surface. These satellites might be equipped with low resolution cameras which would provide a new global cloud map (except for polar areas) every few minutes to any station having the proper receiving equipment and located within "line of sight" of one of the four stationary satellites. Other cameras capable of viewing smaller areas in greater detail also might be included in the satellites. The orientation of these cameras could be controlled by regional forecast centers to study critical areas such as developing storms.

INTERNATIONAL ASPECTS

It is quite clear from the foregoing that the advent of meteorological satellites will have a strong impact on operational and research meteorology throughout the world. In promoting international co-operation in these areas, it is anticipated that the U.N. World Meteorological Organization (WMO) and the International Council of Scientific Unions' Committee on Space Research (COSPAR) will play important roles.

The World Meteorological Organization, affiliated with UNESCO and in existence since 1950, is the lineal descendant of the International Meteorological Organization, established in 1878. The WMO, composed of 108 members, has these purposes: the facilitation of world-wide co-operation in the establishment of observing networks, the promotion of rapid exchanges of weather information, the encouragement of standardization of meteorological observations, the application of meteorology to transportation, agriculture, etc., and the encouragement of research and training in meteorology.

In this framework of international co-operation in meteorology extending back eighty-three years, the appearance of meteorological satellites presents some novel aspects and opportunities. First, in contrast to most conventional observing stations, the satellite observatory is not at a fixed location in the home territory or on the high seas. Earth-orbiting satellites are truly global in their range, and thus enable the meteorologists of the launching country to observe atmospheric phenomena over their area and outside areas more rapidly and more completely than can be done by local meteorologists, and to make some measurements which cannot be done at all by conventional means, no matter how dense the station network is. Since some of the phenomena observed by meteorological satellites may have rapid and serious consequences to the safety of the population and to the economy of a nation, it is imperative that meteorological satellite information be conveyed speedily to all the nations. This can be done in several ways:

Read-out of data by the launching country.—The launching country relays data summaries on international meteorological circuits by means of coded messages or by facsimile. This system of notification introduces delays of several hours which may seriously limit the value of the information.

Read-out of data by individual nations or groups of nations.—This would seem to be the most efficient way of disseminating such information and putting it to immediate use but, as indicated earlier, involves rather expensive equipment of the order of $5 million or more per receiving station and would complicate the satellite instrumentation if all data were to be transmitted in this manner. A more realistic possibility would be the continuous transmission of some of the data (such as cloud pictures over a limited area) with the remaining data being received and processed at a central location before international transmission.

Communications satellites.—Here the launching nation could receive the information from its read-out stations and transmit it in digested and analyzed form to other nations via communications satellites. Raw information from meteorological satellites might also be relayed via communications satellites directly to central processing centers not in line of sight of the meteorological satellites.

A small beginning has been made in limited international release of meteorological satellite information from Tiros II and III. Through the auspices of the WMO, cloud data are transmitted to many countries over a co-operative international meteorological teletypewriter network. Also daily transmissions of cloud charts by United States Navy radio facsimile were beamed abroad for fleet units and received by other nations within range of these transmissions. The WMO informed its member countries of these transmissions and provided pertinent communication information.

WMO can play an important role by keeping its members informed of the launchings of meteorological satellites, their orbits, heights, periods, characteristics of the sensing equipment, frequency and power of the radio transmissions, etc. It could encourage the member countries to participate in the program either by launching meteorological satellites in desirable orbits which supplement those launched by other countries, by assisting in the tracking and receiving of the transmitted data, or by taking valuable auxiliary observations such as additional upper air soundings, and cloud observations. The WMO could also promote the dissemination of such data and stimulate their use and interpretation by distributing literature and arranging for the visits of meteorologists trained in the new techniques or by sending meteorologists to countries which have launched meteorological satellites.

COSPAR, the non-governmental Committee on Space Research organized by the International Council of Scientific Unions (ICSU), which sponsored the International Geophysical Year, could encourage and co-ordinate meteorological satellite experiments and could, because of its broad representation from the international scientific community, draw upon a wealth of knowledge to suggest new meteorological experiments and to promote the design, development, and testing of new observing techniques. These techniques, based on developments in the fields of spectroscopy, electromagnetic radiation, particle dynamics, etc., may depart widely from conventional meteorological instrumentation.

The interplay of the various scientific disciplines engaged in research in space may lead to the unexpected development of valuable new tools and concepts in meteorology. These might include detailed observations of radiations and energetic particles from the sun and space which affect the earth's atmosphere, and observations of atmospheric properties of other planets, as discussed previously.

World Data Centers, containing meteorological satellite data for research purposes by all meteorologists, would also be maintained. World Data Center A, Asheville, North Carolina, is now receiving data from United States meteorological satellites.

In considering a world system of meteorological satellites for both operational and research use, the question of financial support arises. Who would pay for the satellites? Will this be done by individual countries or will there be a sharing of the costs by many countries, with a few countries designated to do the launching? For the foreseeable future, it is likely that costs of the meteorological satellites will be borne by individual launching countries, but other countries might establish their own read-out stations. However, it is interesting that Dr. A. Viaut, president of the WMO, on the occasion of its tenth anniversary, stated: "It would not be realistic to mention—except as an ideal not likely to be attained for a very long time yet—the possibility of a single world meteorological budget." Until this day arrives there is ample opportunity for continued increase of international co-operation in meteorology, and it is expected that meteorological satellites will contribute strongly to this movement.

SUGGESTED READING

BANDEEN, W. R., HANEL, R. A., LICHT, J., STAMPFL, R. A. and STROUD, W. G. "Infrared and Reflected Solar Radiation Measurements from Tiros II Meteorological Satellite," *Journal of Geophysical Research* **66** (1961) 3169–85.

BRISTOR, C. L., and RUZECKI, M.A., "Tiros I Photographs of the

Midwest Storm of April 1, 1960," *Monthly Weather Review,* **88** (1960), 295–314.

FRITZ, S. "Satellite Cloud Pictures of a Cyclone over the Atlantic Ocean," *Quarterly Journal of the Royal Meteorological Society,* **87** (1961), 314–21.

HUBERT, L. F. "A Subtropical Convergence Line of the South Pacific," *Journal of Geophysical Research,* **66** (1961), 797–812.

KRUEGER, A. F., and FRITZ, S. "Cellular Cloud Patterns Revealed by Tiros I," *Tellus,* **13** (1961), 1–7.

NEIL, E. A., and ALLISON, L. J. (eds). "Final Report for the Tiros I Meteorological Satellite System," *NASA Technical Report* No. R–131 (1962).

STERNBERG, S., STROUD, W. G. *et al.* Collection of papers on Tiros I meteorological satellite, *Astronautics,* June 1960.

WHITNEY, L. F., and FRITZ, S. "A Tornado Producing Cloud Pattern Seen from Tiros I," *Bulletin of the American Meteorological Society,* **42** (1961), 603–14.

WINSTON, J. S. "Satellite Pictures of a Cut-off Cyclone over the Eastern Pacific," *Monthly Weather Review,* **88** (1960), 295–314.

WINSTON, J. S., TOURVILLE, L. "Cloud Structure of an Occluded Cyclone over the Gulf of Alaska as Viewed by Tiros I," *Bulletin of the American Meteorological Society,* **42** (1961), 151–65.

SPACE RESEARCH AND THE EARTH SCIENCES

G. P. WOOLLARD

Geodesy is the science which deals with the shape and size of the earth. It provides the skeletal framework for cartography, the visual representation of the earth in the form of maps, and is intimately related to navigation, which is concerned with methods and techniques for positioning and guiding a ship, plane, or missile from one point to another. Geophysics, which deals with all physical phenomena related to the earth, provides the basic data upon which we depend for our knowledge of the internal structure and composition of the earth. The satellite program has had a profound impact upon these diverse fields of knowledge and inquiry as the result of a common element—the earth's gravitational field.

Although gravity is one of the most fundamental phenomena, its nature is far from understood, despite intensive study by many of the world's scientists. The most that can be said at present is that there are two schools of thought regarding gravitation: gravity is purely a geometric phenomenon which involves no radiational transfer of energy between bodies and hence is not subject to modification; or gravity is in part the result of energy-carrying "gravitational waves" which are characterized by energy quanta called "gravitrons." Even under the latter view, the amount of energy transported is believed to be so small as to be negligible; hence any transmutation of gravitrons would have little effect on the gravitational attraction between bodies.

These conclusions are important in satellite studies for several reasons. The numerous bodies in space have gravitational effects on the paths of man-launched space vehicles. There is the possibility of gravitational shielding when a third body is introduced between two bodies in space, and there would be advantages to be gained by using an anti-gravity device in space probes. There is no experimental evidence from either laboratory studies or satellite orbital studies to indicate that gravitational attraction is other than a geometric phenomenon, and it is difficult to understand some Russian statements that vehicles not subject to gravitation are being planned in that country for the exploration of outer space. One can only conclude that such remarks are irresponsible, and certainly some Russian physicists have taken exception to them.

Though the nature of gravitation is not understood, its effect can be accurately evaluated. As long ago as 1670 Sir Isaac Newton showed that the gravitational attraction between two bodies could be expressed as $g = G\ (M_1\ M_2/R^2)$ where G is a gravitational constant, M_1 and M_2 the masses of two bodies, and R the distance between them. Although the value of the constant G was not known then, Newton was able to demonstrate the validity of the gravitational expression in what was, in fact, a pioneer quantitative analytical study of a satellite. Newton reasoned that the moon was held in orbit about the earth by the gravitational attraction of the earth, and should be falling toward the earth by some amount in a given interval of time that was a function of its velocity, gravitational attraction, and the horizontal distance traveled during the same time interval. The basic reasoning Newton used can be simply illustrated: (1) the moon has a nearly circular orbit about the earth at a mean distance of about 240,000 miles from the earth's center; (2) if the acceleration of gravity at the earth's surface (a distance of about 4,000 miles from the center) is 980 cm/sec^2, then at the distance of the moon, it should be $(4,000/240,000)^2$ or $1/3,600$ of 980 cm/sec^2 = 0.272 cm/sec^2; (3) applying this value in the relation $s = \frac{1}{2}\ gt^2$ discovered earlier by Galileo, which relates acceleration of fall (g), distance (s), and time (t) for a freely falling body, the moon's orbit should depart from a straight line because of its attraction toward the earth by 0.136 cm/sec or 10,200 km. (approximately 6,300 miles) per day.

Astronomical observations show that the moon has an orbital period of 27.3 days or 13.2° per day. At a distance of 240,000 miles, the difference in the distance to the orbital path and a line drawn tangent to the orbit 13.2° away from the point of measurement, is a close approximation to 6,300 miles, the distance calculated on the assumption that the moon's orbit is governed by the gravitational attraction of the earth.

Although the orbit of an artificial satellite about the earth is governed by the same forces (gravity and velocity) that determine the orbit of the moon, the fact that the satellite is much closer to the earth than the moon results in its sensing many localized variations in the earth's gravitational field that have lost their individual identity at the distance of the moon. A rough analogy would be the integration of the beams from two flashlights with distance. Close to the source the two beams are distinct and recognizable, but at a distance the illuminated field appears to come from a single source rather than two. As a consequence of this proximity effect, the orbit of an artificial satellite is characterized by deviations in orbital path (perturbations) that can be utilized for studying both the basic shape of the earth, and inhomogeneities in terrestrial mass distribution.

The principal perturbating effect is the change in gravitational attraction between the equator and the poles due to the rotation of the earth and the induced oblateness. Secondary effects are the asymmetry in the distribution of continental masses between the northern and southern hemispheres, changes in mass distribution within the earth, surface and near surface inhomogeneities in mass distribution of geologic origin, and mass effects associated with possible geometric abnormalities, such as the earth's having an elliptical equatorial section. As the observed perturbation of orbit at any one point may be caused by the integrated gravitational effect derived from several unrelated sources, the problem faced in making analytical studies of the observed orbital perturbations is formidable. Further, the reliability of any analytical solution regarding a single factor such as the departure of the earth's shape from that of a sphere will depend not only on how well the various factors affecting the gravitational field sensed in space can be separated and identified but also on how well other perturbating effects related to factors such as atmospheric drag and solar radiation pressure can be

Space Research and the Earth Sciences

determined and taken into consideration.

Analytical studies of the orbital perturbations of artificial satellites suggest that our present knowledge of the gravitational field of the earth is limited. This is not surprising as detailed surface gravity observations have been made over less than 18 per cent of the earth's surface. These orbital studies also suggest that we really do not know the shape of the earth. They suggest that the earth has not only a flattened "tomato shape," which is the basic form that has been accepted and used for all geodetic work since the time of Sir Isaac Newton, but also a superimposed "pear shape." This "pear shape" was derived from a harmonic analysis of orbital perturbation data. It represents a third harmonic term that will undergo some modification with time as reliable values become available for higher order odd number harmonic terms which were not evaluated but incorporated as an integrated effect with the third order term. There is also the possibility that later analyses based on the orbital perturbations of artificial satellites having markedly different inclinations from the ones studied to date may yield a significantly different third harmonic term. It is even conceivable that the "pear shape" may disappear entirely with time. In any event, it must be realized that any model of the earth defined from orbital perturbation data is a dynamic one and related to the earth's mass distribution which may or may not have geometric expression.

In view of the above remarks, why do scientists now pay any serious attention to deductions from satellite orbital observations concerning the shape of the earth? They do so because there is an immediate need for knowing the dynamic form of the earth in connection with many aspects of the space program, and because standard geodetic methods for determining the earth's shape, when applied in different parts of the world, do not yield compatible data. This condition is a result of the fact that the basic measurements required, namely, earth curvature and the distance between intercontinental points, cannot be measured directly, and must be calculated using the celestial positions of heavenly bodies as reference points. The key measurement, that of the angular relationship between a terrestrial point and a celestial point, of necessity must be referred to the direction of gravity as defined by a plumb bob or a level bubble whose position in the gravitational equipotential surface is perpendicular to the vertical defined by the plumb bob. Therefore, any gradient in the gravitational field will tilt the reference system and thus affect any measurement referred to it.

The significance of this effect in key geodetic measurements may be illustrated by, considering first the determination of earth's curvature. The arc measurement technique used is based upon the determination of the interior angle subtended by a measured arc at the earth's surface. Let it be assumed that the shape of the earth approximates a sphere, and that L is the distance between two points which lie on an arc described by the curvature of the earth. The circumference of a medial section of the earth from geometry is $2\pi r$, where r is the radius, and the central angle corresponding to the arc length (L) is $V°$. If L and $V°$ can be measured then r can be solved from the relation $L/2\pi r = V°/360°$. The first such measurement was made by Eratosthenes in 220 B.C. He observed that on the longest day of the year the sun shone vertically down a deep well at noon in Syene in upper Egypt. On the same day at Alexandria, which lies due north of Syene, the sun at noon cast a shadow from a vertical stake driven in the ground. Using the height of the stake and the length of its shadow, Eratosthenes reasoned that the angle defined was related to the amount of earth curvature between Syene and Alexandria, and he was able from his knowledge of the distance between these two points to compute the earth's radius. Eratosthenes' value of r was about 16 per cent greater than that deduced from modern measurements, which is reasonably accurate considering the inexact measurement of L, which was based upon how far a camel caravan traveled per day. Modern measurements differ from those employed by Eratosthenes only in that the astronomical observations are made with precision theodolites and the arc distance is based upon chains of triangles expanded from a base line whose length has been measured with high precision. Despite these refinements, the curvature obtained in any one area may be misleading in regard to the earth as a whole, because the value of the central angle

$V°$ measured is controlled by the gradient of the local gravitational field. A major objective of geodesy is thus to minimize this effect or completely circumvent it. An obvious independent approach to the problem of the earth's shape is to make an analytical study of the earth's gravitational field, since it does change in a more or less systematic manner by 0.5 per cent between the equator and the poles.

Various attempts have been made to do this, but because of the limited amount of data available and their distribution, the poor quality of some of the observations, and the fact that there was no international gravity standard prior to 1960, no geodetic analysis of gravity observations to date can be regarded as really significant.

As a result of all those limitations, there are now seven different geodetic models of the earth in use with varying degrees of oblateness for the earth varying from $f = 1/293.5$ to $f = 1/300.8$, where $f = (a-c)/a$, and f is the flattening, a the equatorial radius, and c the polar radius. It is significant that most analyses made of the orbital perturbations of artificial satellites suggest that the value of f should be $1/298.3$, which differs from the value $f = 1/297.0$ that has been adopted by the International Association of Geodesy as a most probable value on the basis of geodetic and gravitational studies. However, a recent analysis of satellite data (Izsak, 1961, which incidentally shows that the earth has equatorial ellipticity) suggests the value $f = 1/296.9$. The results from satellite studies are therefore not all in agreement and it is clear that the problem of the geometric representation of the earth is far from resolved. These results also emphasize the need for a co-ordinated program of independent studies.

WHERE ARE THE CONTINENTS?

The problem of determining the relative positions of points on the earth's surface is well illustrated on some ocean islands of volcanic origin. There is a high concentration of mass in the central feeder pipes, with the bulk of the island being made up of lower-density ejected material. The resulting distortion of the gravitational equipotential field affects both local positions relative to each other, and the position of the island relative to other places. On Bermuda, for example, there is a quarter-mile discrepancy in a distance of 15 miles between two points as determined astronomically and as determined by actual measurement. It is therefore not surprising that the position of North America relative to Europe may be in error by as much as 400 feet, and that of North America to Asia by up to 1,200 feet. The knowledge of position of other continents, such as Australia and Antarctica may be even more inaccurate.

Fortunately the resolution of these problems of position as well as those of the earth's geometric shape and dynamic form, that appeared to lie well in the future only five years ago, can now be undertaken with every assurance of success. Four independent approaches can be utilized, but preferably they should form parts of a co-ordinated study. These are: (1) A new analysis of observations of gravitational effects based on recent data obtained from a planned program of global observations and incorporating older data adjusted to give the closest agreement with the recently completed network of world standardization measurements. A global program of measurements is now a practical possibility because of the development during the past five years of suitable instrumentation for observations from aircraft and surface ships. (2) An extension of the program of electronic (Hiran) measurements of distance, (*a*) between continents, (*b*) along arcs of measurement employed in earlier geodetic studies, (*c*) connecting the various geodetic datums, and (*d*) along new arcs. (3) Integrated analyses of the perturbations of the orbits of several satellites with different orbital inclinations, with particular emphasis on resolving the value of higher order odd number harmonic terms and the longitude terms. (4) Placing a specially instrumented and designed satellite in an orbit planned to yield a maximum amount of geodetic information.

At first glance a special geodetic satellite might appear to be unnecessary now that so many satellites have been launched whose orbital perturbations can be studied. However, when one considers the various factors that can influence the orbits of these satellites, most of which travel along highly elliptical paths and are subject to variable degrees of atmospheric drag, solar radiation pressure, ionospheric charge, and magnetic field strength, and the fact that few can be sighted simultaneously from widely separated points on the earth's surface or tracked continuously with the requisite precision, it is clear that a special geodetic satellite has much to offer. For this reason the committee on geodesy of the Space Science Board of the United States National Academy of Sciences has recommended that a special satellite be launched to resolve some of the basic problems of geodesy, and at the same time obtain more knowledge of the earth's over-all gravitational field. The following recommendations for this satellite have been made:

The satellite should be spherical in shape, with a high mass-to-volume ratio to minimize atmospheric drag effects. Its orbit should be essentially circular, have a high inclination, and a perigee distance of at least 800 miles from the earth. Such an orbit would eliminate, as nearly as possible, marked variations in perturbating effects of extraterrestrial origin, minimize the effect of atmospheric drag and of changes in density of the medium, keep the satellite high enough to be viewed simultaneously from widely separated points on the earth, and also keep it visible for optical tracking from any one point for a maximum period of time. The satellite should be equipped with a flashing light source that could be coded and flashed on command so that simultaneous observations could be assured for groups of observers at widely separated places. The satellite should be equipped with a radio beacon transponder system so that electronic range measurements and bearings can be obtained from fixed ground stations as well as optical bearings. This would also assure tracking during periods of poor visibility because of bad weather. The satellite should have a solar battery power supply to assure maximum longevity of use, and corner reflectors so that it can be tracked optically if instruments or power fail.

Such a satellite, with a judiciously located network of ground tracking stations, would yield data not only on the earth's gravitational field and the size and shape of the earth, but also precise information on the distances between existing geodetic datums and isolated triangulation networks in different parts of the world. It is believed that position control between continents using a geodetic satellite can be accomplished with a precision of about 30 feet. This is the same order of accuracy that is achieved over relatively short distances with electronic distance measuring systems such as Hiran. Hiran (high precision Shoran), an outgrowth of the navigational system Shoran (short-range navigation), is mentioned specifically because it is the electronic surveying system that has been used successfully to establish intercontinental connections and to establish the length of long arcs of measurement up to $100°$. An aircraft equipped with a Shoran transmitter and receiver flies across the line whose length is to be measured. Every two seconds a radio frequency pulse signal is sent out which is received and retransmitted back to the plane by transponders located at each end of the line. The time lapse between the origin of the signal and its return from the ground stations is automatically converted to distance and thus the slant range to each station determined. At some point in crossing the line the sum of the distances to each ground station will reach a minimum value, and it is at this point that the aircraft is in the vertical plane connecting the two ground stations. By making a series of such crossings at two fixed altitudes it is possible, with certain corrections to convert the minimum-sum distance values to the equivalent arc distance between the two points. Such measurements have been made between Europe and Africa, across the Mediterranean Sea, and between North America and northern Europe via Greenland and Iceland. Others are now being made. However, because the longest single connection of this type made to date is about 550 miles, it is clear that Hiran will never be able to accomplish intercontinental ties across broad oceanic reaches where there are no island steppingstones, and as a result the Hiran program can only contribute to the resolution of the basic problems of geodesy, not solve them completely.

Another method of positioning the continents relative to each other, and also obtaining other geodetic data, by using a moon camera as was done by William Markowitz in 1958. Basically this apparatus is a dual-rate camera which can be attached to either a visual or photographic refractor telescope. The camera simultaneously tracks the moon and the surrounding stellar background with the moon held fixed relative to the stars during an exposure, using a tilting internal prism. The developed photographic plates are measured with a precision measuring machine to determine the x and y co-ordinates of about 30 points on the limb of the moon relative to a number of reference background stars, and the position of the center of the moon relative to the positions of the stars is then determined by a least-square solution.

Dynamic information on the size and shape of the earth can be obtained because the parallax of the moon is such that an observer on the earth sees the moon displaced from its astronomical position by an amount which depends upon the observer's position on the earth and the distance of the observing site from the earth's center. By combining data from observations made at several places around the earth, it is then possible to delineate the moon's orbital path and perturbations of the orbit, which in turn can be related to the size and shape of the earth and abnormalities in its gravitational field. By using the moon as the third point of a triangle whose altitude is known, it is possible to determine intercontinental distances geometrically by angular measurements from two points on the earth's surface. The accuracy of the method for such connections is believed to be about 150 feet.

Although the basic data to be obtained in using the moon for geodetic studies of the earth are the same as those to be obtained with an artificial satellite, there are distinct advantages to be gained from an artificial satellite. It is possible to put instruments in the satellite that will improve the precision of the tracking capability; an artificial satellite can be placed in any preselected orbit to yield a maximum amount of data bearing on a specific problem; and more than one satellite can be used to obtain supplementary information requiring a different orbital inclination or perigee distance.

A third method of untested reliability for determining intercontinental connections is a sonic one based on the time it takes a sound pulse to travel between continents via the deep sound channel in the oceans. This system, known as Sofar, depends upon originating and receiving a sound signal at the depth of minimum velocity in the oceans, and the velocity of sound transmission at the axis of the sound channel remaining constant. Such signals, generated by a small charge of explosive (5 to 20 pounds), have been recorded over distances up to 12,000 miles (Australia to Bermuda). Over restricted ranges where the requirement for constant velocity can be satisfied by the internal temperature and salinity structure within the ocean, it is believed that this system may provide distance measurements as reliable as those believed possible with the moon camera.

None of the systems considered for intercontinental connections has both the general application and the degree of accuracy believed obtainable with a special geodetic satellite, and none has the geodetic satellite's versatility because of a selected orbit for resolving other problems of the earth's gravitational field and geodesy. General applications and versatility can be illustrated in part by the objectives of the ground tracking network. A tracking network performs two operational, scientific services. The space program to date has served two geodetic purposes: it has focused attention on the need for better geodetic knowledge, and it has helped formulate new models of the earth to be tested in other investigations. It has not solved the problems of geodesy. A combined geodetic satellite and gravitational program, however, could come close to doing this. The results from each program would support the other, and both approaches should also give results that are compatible with astronomical data and with geodetic mensuration studies of the geometric form of the earth. A fully co-ordinated study would result in a "unique" solution. We are now on the threshold of a new era in geodesy, an era concerned not only with the geometric representation of the earth, but also its dynamic form as expressed by the earth's gravitational field in space.

NAVIGATION AND TRACKING

Just as gravity, geodesy, and the orbit of a satellite are intimately connected, so are gravity, navigation, and tracking.

Both navigation and tracking involve positioning a body in some reference system of spatial co-ordinates. On an earth-bound system, the co-ordinates are latitude and longitude. In a space system, three co-ordinates of position (x, y, and z) are required. The x axis by convention is the direction of the vernal equinox, the z axis is directed toward the north celestial pole, and the y axis is directed so as to form a right-handed orthogonal. The direction of the vernal equinox is also used as the reference zero celestial meridian just as Greenwich, England, is used as a reference zero meridian on earth for longitude. All other meridians are defined in terms of the sidereal hour angle west of the equinox position. The vernal equinox is the point on the celestial sphere where the apparent orbit of the sun crosses the celestial equator from south to north at the time of equal length of day and night (equinox). The celestial sphere is simply an imaginary sphere at an infinite distance from the earth. Because of precession effects, any position in a celestial equatorial system of co-ordinates must be referred to the vernal equinox of a specific year. Latitude on the celestial sphere is referred to as declination, and is designated as north or south of the equator as on the earth. In celestial navigation the stars, planets, moon, and sun are used as heavenly reference points whose positions relative to points on the earth have been established as a function of date and time. To determine position by celestial navigation, a good chronometer, a sextant, and an up-to-date copy of a star almanac listing the declination and sidereal hour angle of stars commonly used for reference purposes are required.

Gravity enters into the determination of position in celestial navigation when the sextant—an optical instrument for measuring the elevation of the reference body above the horizon when at sea, or above the gravitational equipotential surface defined by a level bubble when in the air—is used. Because both the sea surface and a level bubble are aligned in the earth's equipotential gravitational field, any warping of this surface influences astronomically determined positions. The magnitude of error that can be related to the local gravitational field can be gauged from the fact that an error of one minute in elevation of a star will result in an error of one mile in position.

Optical tracking of an artificial satellite using a gravitationally controlled reference will be biased by the local gravitational field at a tracking station in the same degree as its own astronomically determined position is biased. Allowance can be made for the local warping of the gravitational field, which is defined as a departure of the vertical relative to the local geodetic datum from comparisons of astronomically and geodetically determined positions. Consideration of this factor is not enough, however, because the unknown relationships between the various geodetic datums still leave a degree of uncertainty as to the relative positions of the tracking stations. One objective of the proposed geodetic satellite program is the resolution of these uncertainties through precise simultaneous bearing and electronic distance measurements to the satellite from several points located on different geodetic datums.

Electronic methods of determining bearing and distance are currently being utilized in satellite tracking, and would have a key role in exploiting to the fullest extent the scientific potential of a geodetic satellite. Electronic distance systems can be divided into two groups: those that measure distance directly in terms of the travel time of a reflected or retransmitted electromagnetic pulse (referred to as circular methods); and those that measure differences in distance to two or more fixed points, as indicated by the time delay or phase shift in the signal received from the fixed points. The latter methods are called hyperbolic methods because the fixed points are foci of a series of hyperbolic curves along which there is a constant difference in travel time (distance). Both methods have a history of many years of use in marine navigation as a means of establishing position. The circular methods give data for establishing position on the basis of the intersection of two arcs whose radii correspond to the distances from two fixed points at the ends of a known base line. Shoran, which was developed prior to World War II, is an example of such a system.

Loran (long-range navigation), on the other hand, uses the hyperbolic method of distance determination. In using Loran, the location of an unknown point is determined graphically from special charts on the basis of the intersections of pre-plotted travel time difference curves for a series of fixed Loran stations along the coast covered by the chart. Determining which of the hyperbolic curves on the chart are to be used is done by using synchronized signals from three shore Loran stations and measuring the difference in times of arrival of the signals from each pair of stations at the ship whose position is desired. Such a navigational system can also become a tracking system by using a transponder on the ship and the difference in times of arrivals of the signals retransmitted back to shore for each pair of stations. This change in roles is also illustrated by the Hiran system of distance measurement, which is actually a reversed application of Shoran. Radar, which is used for both navigation and tracking, involves only the accurate determination of travel time for an electromagnetic pulse that has been passively reflected from a target back to the point of origin.

The electronic system used for tracking artificial satellites (Minitrack), although technically a hyperbolic distance measuring system, is actually used principally to establish the line of position between the satellite and the tracking station. Two perpendicular base lines are employed, and at the ends of each base line there are receiving antennae. A radio signal from a satellite in space will be picked up by the receiver antenna at one end of one of the base lines ahead and out of phase with the signal received at the other end of the line. The difference in phase is a measure of the difference in travel path from the two ends of the base line to the satellite, and is related to both the angle of incidence of the arriving signal and the bearing of the satellite from the azimuth of the base line. From the phase difference of the signal received at the two ends of the base line, the wave length of the radio signal and the length of the base line, a bearing vector is obtained. Similar data derived from the pair of receiving stations located at the ends of the second base line, oriented at right angles to the first line, define the direction of the satellite from the station. This system is capable of measuring angles to 20 seconds of arc. As presently set up there is a "picket fence" alignment of seven such stations along the 75° west meridian, with five other stations located strategically elsewhere. Such an array could obviously also be used as a navigational aid for positioning manned space vehicles in the future in much the same way that Loran is now employed for positioning vessels at sea.

Present tracking facilities also utilize a second phase comparison system, Microlock, to supplement the Minitrack observations. This system utilizes a three antennae system. One antenna supplies the input to a phase-locked receiver designed to detect the Doppler shift (change in frequency) of the satellite radio beacon signal as the satellite advances toward or retreats from the tracking station. Although it lacks the accuracy of Minitrack for determining bearings, Microlock does serve as a communications system for telemetered data, and by using Doppler shift observations, it is possible to determine the range of the satellite and to derive its orbit. There is a limitation on the reliability of electronic distance measurements in space, because of ionosphere refraction effects that are not fixed but fluctuate with solar activity.

The most accurate tracking method for determining the orbit of a satellite is provided by the Baker-Nunn optical photographic method. This system uses time-co-ordinated measurements referred to a background of fixed stars, as does the moon camera. The tracking capability of the Baker-Nunn cameras is phenomenal. An object as small as a 6-inch sphere (1958 Beta 2) has been tracked to distances of over 4,000 kilometers (2,400 miles). The cameras take successive exposures at intervals of from 2 to 23 seconds, have a field of 5° by 30°, and can photograph an image as faint as 13th magnitude. Accuracies in position determination are believed to be better than 2 seconds of arc normal to the satellite trail and 0.01 second of time along the trail. At a distance of 200 miles these figures correspond to a position accuracy of about 10 feet normal to the satellite trail and 25 feet along the trail. The only limitation of this method is that imposed by atmospheric and light conditions. The use of a star background reference system eliminates the effect of the local gravitational field, and although the relative positions of all stations are not accurately known this has not proved to be critical. Data from the Baker-Nunn tracking stations, incorporated with accurate time measurements, made it possible to establish tables of orbital elements for various satellites with sufficient precision to permit their use as celestial navigational aids. More important than this potential use of artificial satellites for navigational purposes, however, is the use of special satellites for ground position fixing. In less than five years sextants may well be relegated to museums as curiosities like the astrolabe of Columbus' day. To get a "fix" on his position, a ship's navigator will simply "tune in" on one of several navigational satellites and determine electronically his distance from the satellite, and from the known celestial position of the satellite for the date and time of the "fix," determine a line of position. Similar data from three such navigational satellites would provide a "fix" accurate to a few feet. Such a system would not only avoid the errors that are now introduced by vagaries in the earth's gravitational field but would also provide an all-weather system that would permit accurate determination of position at any time of day or night. In addition to position control, such navigational satellites would also provide a means for determining ship's speed, course direction, direction of the vertical, and true north.

GUIDANCE SYSTEMS

A secondary effect of the space program on navigation is the application of the sophisticated guidance systems that have been developed for controlling the trajectories of unmanned space vehicles. These mainly represent outgrowths of basic systems developed during World War II to guide bombers on attack missions and submarines on submerged patrols which were out of contact with normal navigational aids. The principal guidance systems are: homing systems, command systems, beam-rider systems, inertial systems, and power systems. Of these, the inertial systems, which utilize the axial stability of a gyroscope and its constant angular momentum, have been used most frequently in the space program. The gyroscope can be used not only to maintain orientation and to sense relative motion in a given reference system, but also to establish a space reference from which missile motion or position can be measured.

Three of the principal inertial systems developed for missile guidance are: (1) Inertial guidance—the trajectory azimuth along a predetermined path is controlled by a manually pre-set gyroscope with corrections in initial heading to take care of changes in gravity and wind. (2) Inertial gravitational guidance—in addition to ordinary inertial guidance there is a sensor to measure and relate the vertical (direction of gravitational field) at any point along the trajectory path to the vertical at the target. When the two coincide, the missile is on target. This system requires a knowledge of the vertical at the target and that this value be unique. (3) Inertial-celestial guidance—a predetermined course in space related to the relative position of the missile and certain preselected celestial bodies is followed. This system also requires measurements of the direction of the vertical to relate angular positions of the stars tracked to comparative earth co-ordinates. All these systems are terrestrially oriented, with the earth's gravitational field supplying a vertical reference datum.

For a space probe going outside the earth's gravitational field, guidance has to be accomplished in three steps: initial guidance under power that carries the space probe through the earth's gravitational field; intermediate guidance while between the earth and target; and terminal guidance while in the gravitational field of the target. In initial guidance, velocity, the time of burnout, and heading are critical. Once in free flight between the earth and target, altitude control can be maintained by firing small auxiliary gas rockets to create a torque by expelling material, by application of the law of the conservation of angular momentum, and possibly by using solar radiation pressure. This last, however, has not yet been developed into an operational system. Terminal guidance while in the gravitational field of the target can be gravitationally controlled, rocket controlled, or based upon some special sensor such as infrared radiation. The critical guidance phase is the first, when the space probe is put on a trajectory to intersect the orbital path of the target. This requires not only a precise trajectory design but the selection of an optimum date for launching. The magnitude of the problem is illustrated by the fact that even with a miss of 6,000 miles, which would still allow a space missile to be captured by the gravitational field of Mars, the initial heading for certain trajectories would have to be accurate to $0.001°$ and the burnout velocity to one foot per second. Space navigation thus poses problems that are unique to the space environment. Its principal impact on terrestrial navigation will be the automatic guidance systems developed.

Although automatic inertial guidance systems using a gravitationally oriented reference have been widely applied in terrestrial navigation, systems based on the conservation of angular momentum have not been used so extensively. The principle of this method is well illustrated by the change in speed of rotation achieved by an ice figure skater in a spin. A skater is able to speed up his rate of spin by holding his arms close to his body, and to slow down the rate of spin by extending his arms. This is because the angular moment of inertia of a freely revolving mass system (whether an ice skater, a gyroscope, or the earth rotating on its axis in space) has a constant value determined by the rate of rotation and the distribution of mass relative to the spin axis. Any change in the distance relationship of the mass to the spin axis automatically is reflected in the speed of rotation.

To apply this to space vehicle guidance in free flight, consider a wheel mounted in the vehicle with its axle perpendicular to the long axis of the vehicle. If neither the vehicle nor the wheel is rotating, the angular momentum of the wheel-vehicle system is zero. If the wheel is caused to rotate, then it has angular momentum. Because of the principle of the conservation of angular momentum, however, the angular momentum of the wheel-vehicle system cannot depart from the original zero value, and as a result the vehicle reacts to the motion of the wheel and rotates an equivalent amount in the opposite direction. Three such wheels in a vehicle with mutually perpendicular spin axes can thus control pitch, roll, and yaw by commands from gyroscope sensing elements. In this sense outer space guidance is similar to that of an automatic pilot system on an aircraft. The difference lies in the control system actuated by the sensing system. Developments such as spacecraft guidance systems and the coming of navigational satellites in the near future can thus be expected to have a profound effect upon navigation. If geodesy stands on the threshold of a new era, navigation stands on the threshold of a new age—the age of space.

OTHER BENEFITS

The space program is having important indirect side effects in addition to the key roles it is playing in space research, in resolving the problems of geodesy, and revolutionizing navigation. One of these side effects concerns our knowledge of the interior of the earth, its strength, the interrelation of gravitational and magnetic field abnormalities, and possible circulatory or convectional movement within the core. The model of the earth deduced from satellite orbital data to date suggests that the earth has an elliptical cross-section at the equator, a bulge in the southern hemisphere, and an axial length from the equator through the north pole longer than that through the south pole. Although this model of the earth is a dynamic one which may or may not have geometric expression, it does imply that the earth has considerable strength, since such a form departs from the symmetrical biaxial ellipsoid of revolution that would be formed by a rotating fluid body. It implies that the accepted concept of isostatic equilibrium, whereby regional variations in surface mass distribution are compensated by deficiencies in mass at depth, may not actually exist.

These results raise a doubt as to accepted ideas and also present a challenge, a challenge to prove or disprove the implications of the satellite studies. One way to meet this challenge is to make observations of gravitational effects in the key locations where abnormalities in mass are indicated. For example, if there is an ellipticity at the equator of 410 meters with a semi-major axis in the vicinity of Recife, Brazil, as has been suggested recently, a program of gravitational observations in Brazil and along the northern coast of New Guinea and in Somaliland should indicate the general validity of the model. If the model is substantiated, a second step would be to supplement observations of gravitational effects both at sea and on land to develop the degree of reliability of the model. A third step would be to set up other studies, particularly seismological studies, to see to what extent the composition and structure of the crust and the underlying mantle contribute to the gravitational effect. Another study would involve the determination of whether the dynamic undulations are related in any way to the areas of magnetic field abnormality; whether there is any correlation with heat flow or abnormal thermal gradient. There is nothing new about these studies, but there has never been a completely integrated program of geophysical study of the earth's interior. One reason has been the lack of a specific reason to study a particular area rather than another. The artificial satellite program is now supplying such an objective in pointing to specific areas that may be different in some way. The challenge is irresistible, and the outcome can only be beneficial. It will either lead to new and possibly spectacular knowledge of the internal structure of the earth or result in a critical reappraisal of the analytical methods now being used in studying the orbital perturbations of artificial satellites. Regardless of the outcome, science will have advanced, needless confusion about the shape of the earth will be eliminated, and a new era of co-ordinated geophysical studies of the earth's interior initiated.

SUGGESTED READING

Izsak, I. G., "A determination of the ellipticity of the earth's equator from the motion of two satellites," Smithsonian Institution Astrophysical Observatory Research and Space Science, Special Report 56 (1961).

Ivanenko, D. D., "The enigma of gravitation," *Tekhnika molodezhi* No. 5, p. 11–13 (1959).

Markowitz, W., "Photographic determination of the Moon's position and applications to the measure of time, rotation of the earth, and geodesy," *Astronautical Jour.*, 59 (1958), No. 2, 70–72.

Stanyakovich, K. P., "Scientists predict new 1958 discovery," Tass Foreign Broadcasts Information Service, Moscow, December 31, 1957. Also *New York Times* article of March 2, 1958.

COMMUNICATIONS SATELLITE SYSTEMS

LEONARD JAFFE

We are presently on the eve of an important advance in the art of communications. Since early discoveries of the propagation of electromagnetic radiation through space, and the applications of this to the transmission of intelligible signals, we have learned to make most sophisticated use of radio transmission. The aural senses of man have been extended by radio, and his visual senses have been enhanced by television. Until now, however, the distances over which high-capacity radio transmissions are effective have been limited. High-frequency radio, utilizing ionospheric reflection, has never been a satisfactory medium. In order to achieve the bandwidths required for today's complex communications, we have had to move higher and higher in frequency where the horizon limits the range of radio communications, and we have had to rely on coaxial cables or microwave systems for long-distance transmission of a large volume of information. There exists no medium today capable of transmitting a television bandwith across the great oceans.

Artificial earth satellites offer the possibility of placing high-capacity microwave relays at such distances above the earth's surface that horizons seen by these satellite relays will encompass entire continents and the shores of continents on both sides of the oceans. For the first time, we can visualize communications systems capable of reaching all areas of the earth, with the reliability and capacity required by the demands of today's communications techniques.

It should be understood that communications satellite systems currently contemplated are for point-to-point intercontinental communications. That is, they will serve as long-distance, high-capacity trunk lines, just as microwave relay systems do today.

These systems will not broadcast with sufficient power for reception by home television or radio receivers. Satellites capable of broadcasting to home receivers have been predicted, but this is many years away because of the power requirements of such a service. Let us examine this further. This kind of service would most certainly require a stationary or 24-hour orbit satellite so that the receiving antenna could be pointed in a fixed direction. Even if one assumed a 21-decibel ground receiving antenna attached to a good television set, the radiated power required from the satellite for primary television service is approximately 250 kilowatts at 470 megacycles per second, if near-hemispheric coverage is desired with a single satellite. Approximately 10 kilowatts of radiated power would be required for radio broadcasts (FM at 100 megacycles) from a synchronous satellite. Such power levels will probably not be available aboard satellites for many years.

TYPES OF COMMUNICATIONS SATELLITES

There are many types of satellites which can be used for communications, but they can be separated into two broad categories. Passive satellites are those which merely reflect radio waves, which are beamed at the satellites, back toward the earth. Because all of the energy for communication must originate at the ground transmitter, passive satellites generally require the use of high-power transmitters, large antennae, and very sensitive ground receivers. Passive satellites may be made in many different shapes, but in general they have one thing in common. They must all be large, to present as large a reflecting cross-section as possible, and they must be light, to permit this large reflector to be put in orbit with rocket boosters of reasonable size.

The other category of communications satellites consists of the active satellites. Active satellites contain electronic components—a receiver for signals from the ground, a transmitter to amplify these signals and rebroadcast them back to the ground, and the power supplies necessary to operate these components. Because active satellites amplify the signals transmitted, smaller

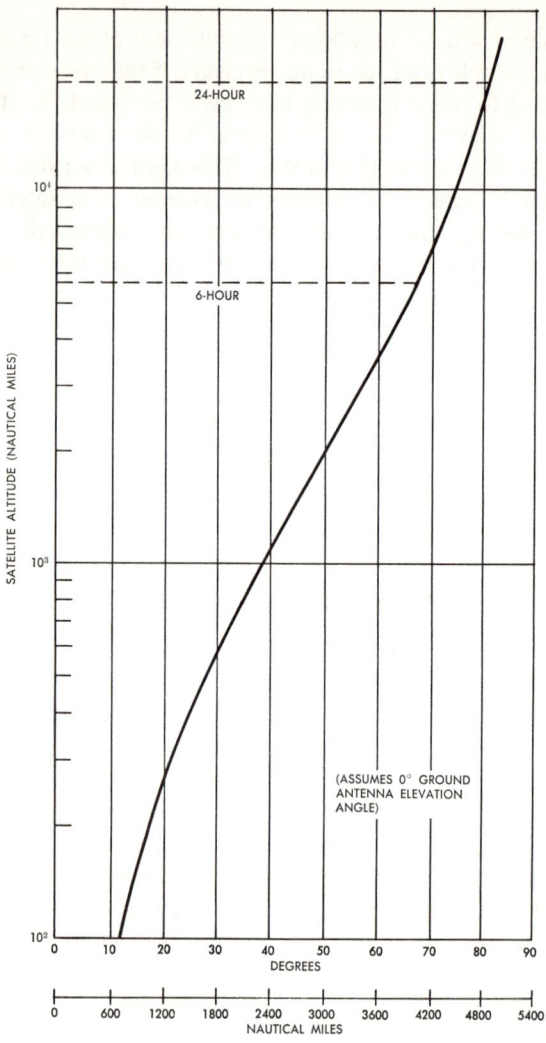

Fig. 1—Communications satellite altitude plotted against ground coverage in nautical miles.

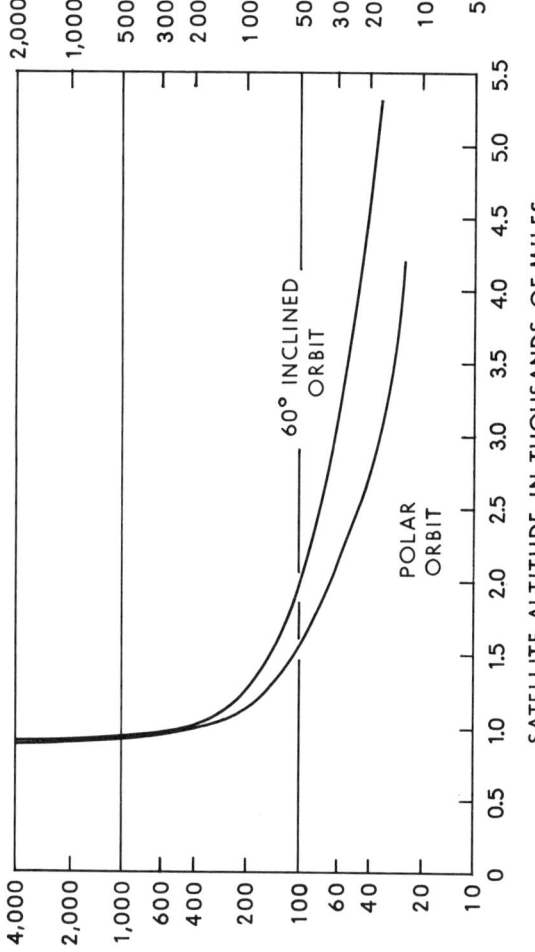

FIG. 2—Number of satellites required for North Atlantic communications. Polar and 60° orbits are shown in thousands of miles; left side of graph represents number of satellites required to provide a circuit 99 per cent of the time; right side shows number of 90 per cent satellite coverage.

ground transmitters and antennae can be employed than with passive satellites.

The advantages and disadvantages of each type are obvious. The active satellite requires smaller ground installations, while the passive satellite has no electronic components on board which might fail and render it useless. The passive satellite is available to many users at the same time, while this kind of accessibility must be deliberately designed into an active satellite.

There is no doubt that the active satellite today is enjoying the most interest from commercial users because of the relative ease of obtaining large bandwidth communications with nominal-sized ground installations.

Many orbital configurations have been proposed for communications satellites. If "real time" communication is desired (and there is universal agreement that it is) and transoceanic earth distances are to be spanned by such systems, the minimum orbital altitude is approximately 2,000 miles. The orbital altitude of the system will be a compromise between conflicting factors. The higher the satellite, the longer the period of mutual visibility between communication sites, and the greater the extent of earth coverage by a single satellite (Fig. 1). The number of satellites required to provide essentially continuous communication varies inversely with altitude (Fig. 2). Several factors operate against the use of higher altitudes. The power required for communication, the time delay of the transmission path, and the cost of booster rocket vehicles to place the system in orbit, all increase with altitude. The influence of these factors will be treated in the discussion of the various systems.

PASSIVE COMMUNICATION SATELLITES

Of the many shapes which have been proposed for passive satellites, the sphere, earth-oriented partial spheres, and orbiting dipoles show the most promise. Spheres of the Echo I (Fig. 3) type reflect impinging radio waves in a non-directional manner and, since they are symmetrical, do not impose a requirement for orienting the satellite continuously in a given direction. Because of this very advantage, the sphere is an inefficient reflector.

FIG. 3—National Aeronautics and Space Administration's Echo I satellite. Echo I is a rigid, inflated sphere, 100 feet in diameter. It has been used to reflect radio and radar signals in "passive" communications experiments. The reflecting surface of the sphere is a very thin film of plastic coated with a vacuum-deposited film of aluminum.

Large amounts of energy are reflected in undesirable directions, making necessary the use of rather high-powered ground transmitters. The moon is really the first spherical passive communications satellite, and it has been used successfully for several years by the United States Navy. There is only one moon, however, and unfortunately it is available for communications to a given geographical area for only about six hours a day.

The communication capacity for a system employing spherical passive satellites is given by: Bandwidth = $P_T G_T G_R \lambda^2 \sigma / (4\pi)^3 R^4 K T (S/N)$, where P_T = transmitted power, G_T = gain of transmitting antenna, G_R = gain of receiving antenna, λ = wavelength, σ = cross-section area of satellite, R = distance from transmitter to satellite, K = Boltzmann's constant, T = noise temperature of receiver, and S/N = signal-to-noise ratio desired. In passive systems, the capacity is a function of the product of the gains of ground antennae $G_T G_R$. This means that the two antennae need not be of the same size. All other factors being the same, a system employing two different-sized antennae would behave the same in either direction of transmission.

There is a possible system, which has been outlined by J. R. Pierce, which would make use of both the moon and man-made passive satellites. A ground complex is suggested, with the following parameters:

	Small Terminal	*Large Terminal*
Transmitter power	4 kilowatts	40 kilowatts
Frequency	4 gigacycles	4 gigacycles
Antenna diameter	7 feet	68 feet
Receiver noise temperature	300° Kelvin	30° Kelvin

Using the moon as a reflector, such a set of stations would yield 2,000 cycles per second bandwidth, with a signal-to-noise ratio of 22 decibels, using the small terminal as a receiver, or 32 decibels using the large terminal as the receiver. Using a 140-foot diameter satellite of the Echo II type now being developed by the National Aeronautics and Space Administration at an altitude of 1,700 miles, these stations would provide over 500 cycles per second bandwith in either direction, with a signal-to-noise ratio of 21 decibels.

If two of the larger facilities were used to communicate via the 140-foot diameter satellite, a 21-decibel signal-to-noise ratio would result in a bandwidth of 500,000 cycles per second (500 kilocycles). New receiver technology, such as frequency modulation (FM) with feedback could result in much lower carrier-to-noise ratios required for reasonable performance of a circuit. These techniques could extend the bandwidth capabilities of a moon-satellite system to perhaps three megacycles.

Passive satellites which might provide some reflective gain without requiring orientation do not look particularly promising now. With or without stabilization of attitude, if we consider communications with the radio horizons of the satellite and all points in between, gains of no more than perhaps 7 decibels are tolerable and very large cross-sections will always be necessary. Passive orientation controls, employing perhaps the gravity field gradient existent at the satellite altitudes, may provide a means for obtaining and realizing the 7-decibel reflective gain with passive satellites.

A passive system, employing a form of chaff, specifically, orbiting dipoles, has been proposed by W. E. Morrow, Jr., of the Lincoln Laboratories of the Massachusetts Institute of Technology. Now designated Project West Ford, the proposed system makes use of scattering microwave energy from a large number of dipoles (Fig. 4) distributed in orbit about the earth. This technique has the advantage that only two orbiting belts of scatterers, one polar and one equatorial (as shown in Fig. 5), are required to achieve global coverage. The radio propagation properties of this type of system depend chiefly on the spread of velocities of various scatterers, and the bandwidth of the antennae used. When antennae having bandwidths of a fraction of a degree are used, the predicted multipath delay is of the order of 100 to 300 microseconds. This indicates that incoherent modulation-demodulation techniques, which do not require detailed knowledge of the phase of the received signals, should be used. It is estimated that with 100 kilograms of microwave dipoles in a circular orbit of a few thousand miles in altitude, communications capacities of up to a few tens of kilobits per second can be obtained over super-high-frequency circuits employing large parabolic antennae and low-noise maser receivers.

In general, passive satellite systems will have much less channel capacity than active satellite systems with similar ground stations for comparable weights in orbit. However, for limited capacity services, and because of their availability to large numbers of users and their inherent simplicity, passive satellites may be used as an adjunct to large-capacity active satellite systems.

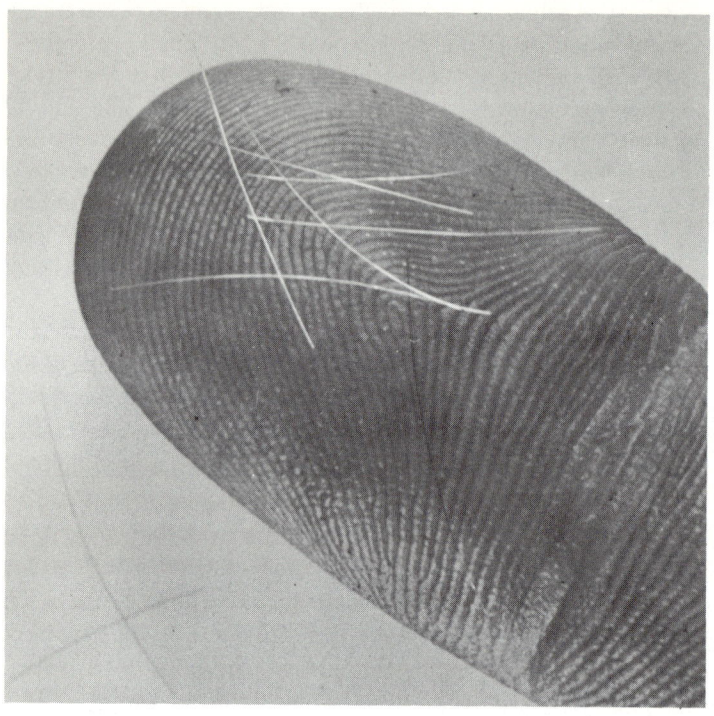

Fig. 4—Orbiting dipole magnets are tiny wires 0.7 inch long. Project West Ford, sponsored by the Air Force, and conducted by Massachusetts Institute of Technology's Lincoln Laboratory, attempted to orbit a belt of such dipoles, which would reflect radio signals, last year; the experiment failed when the material holding the dipoles together failed to sublime, and the belt did not form. (Photo from W. E. Morrow, Jr., and D. C. McLellan, "Proposed Orbiting Dipole Belts," *Astron. Jour.* 66, No. 3, 1961.)

ACTIVE COMMUNICATION SATELLITES

Various systems employing active satellites have been proposed, but they too can be broadly separated into two categories. There is the 24-hour active satellite system (Fig. 6), which consists of satellites placed in a 24-hour synchronous orbit at an altitude of 22,300 miles. These remain stationary with respect to a point on the earth's surface, because the orbital period of the satellite is exactly 24 hours. Because of this unique orbital char-

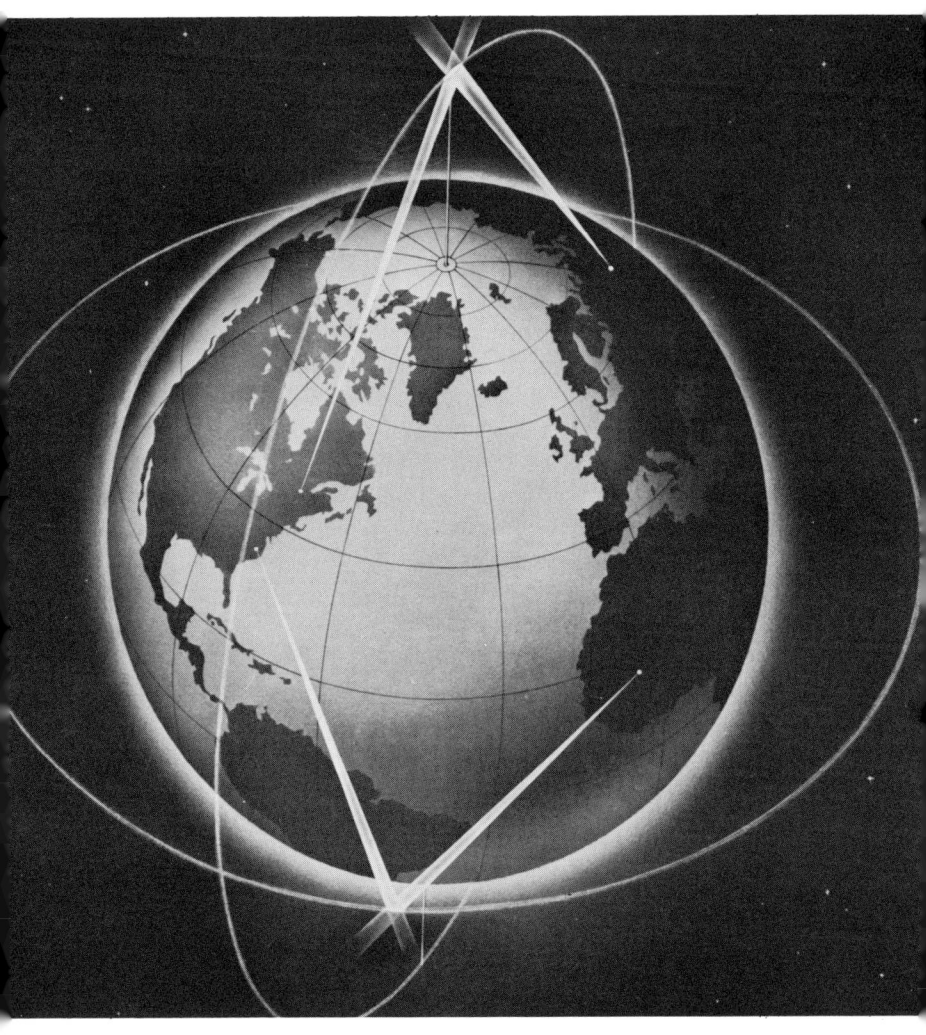

FIG. 5—Orbiting dipole system. Two belts would reflect radio signals as shown. (Photo from Morrow and McLellan, *op. cit.*)

acteristic, as few as three satellites could provide global coverage (with the exception of the polar regions). Such a system has a further advantage in that antenna-pointing problems on the ground are much reduced over lower altitude systems where the satellite's position in relation to the ground stations is continuously changing.

The satellite for the 24-hour system is comparatively complex. Not only must it carry the active electronics and power supplies required for communications, but it must also carry a velocity control system to adjust the position of the satellite periodically, so that it does indeed remain fixed with respect to the earth's surface. Because of the large distance from the earth, satellite antennae must be pointed continuously toward the earth to minimize the power needed to communicate over this distance. This requires an attitude control system. Both the attitude and velocity control systems require that a supply of propellant be carried aboard the satellites. It is generally considered that these control systems will limit the life expectancy and reliability of such satellites. The time required for a radio wave to travel over the 22,300-mile distance and back may cause objectionable delay in a two-way telephone conversation. (The round-trip time delay incurred in a single satellite circuit is approximately 0.6 second).

The other category of active satellite systems can be referred to as low altitude systems (3,000 to 6,000 miles). Low altitude systems (Fig. 7) would necessarily consist of larger numbers of satellites, but they also remove the question of time delay and the requirements for attitude and velocity control systems. Thus, the satellites for such systems can be considerably smaller and lighter, and consequently amenable to multiple-satellite launching, making it possible to place many in orbit with a single booster of the type required to place one satellite in the 24-hour orbit. The primary requirements for commercially operational communications satellite systems are long life and reliability. It is believed that the simpler low altitude active satellite can provide these desired characteristics of long life and reliability on a much earlier time scale than the more complicated 24-hour satellite system. This is a period of research

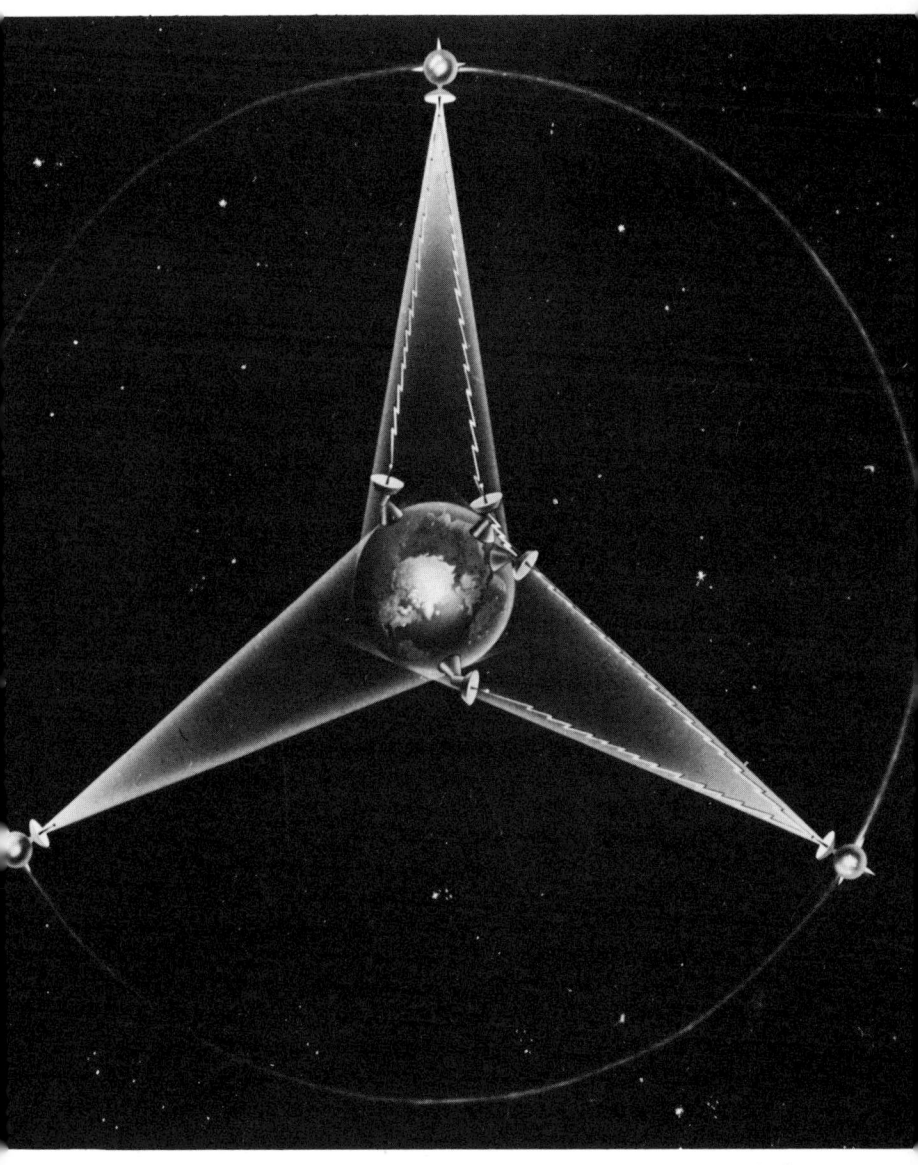

Fig. 6—Synchronous 24-hour repeater satellite communication system. Three such satellites provide nearly global coverage.

and development, and during this period it will be necessary to explore all the approaches.

EXPERIMENTAL FLIGHT PROGRAMS

The experience with communications satellites today is meager. Project Score led to the first active repeater satellite. It was launched and carried the President's Christmas message in December, 1958. The Score satellite was battery powered and performed well during the 30-day life of its batteries.

In August, 1960, Echo I, the 100-foot diameter spherical passive satellite was placed in orbit. It is still in orbit, and although somewhat distorted, it can still be used as a reflector of radio waves. Considerable orbital perturbation data have been obtained from Echo I. Atmospheric pressure at 1,000 miles altitude, the effect of solar disturbances on the density of the atmosphere, and the effect of solar radiation pressure have been measured.

The Courier satellite, a rather sophisticated delayed repeater satellite, was placed in orbit in October, 1960. A delayed repeater satellite is one which receives and stores a message temporarily for rebroadcasting back to earth upon command at some later time during its orbit about the earth. Courier worked well for seventeen days when an undetermined malfunction prevented its further use.

This is the experience, and there are many questions which need to be answered. Space poses many problems beyond just getting a satellite there. One of the serious questions is the detrimental effect of high energy protons on solid state components in the inner Van Allen belt. Most electronic devices can be shielded, but solar cells must be exposed to the environment and cannot be protected. The lower Van Allen belt extends from approximately 600 miles altitude to above 5,000 miles. The upper edge of the belt is not well defined and the energy distribution of particles in the belt needs better definition. Estimates of the life of solar cells in the heart of the belt vary today from a few months to a few years, even when protected by a reasonable amount of quartz or sapphire. The exact nature of the proton

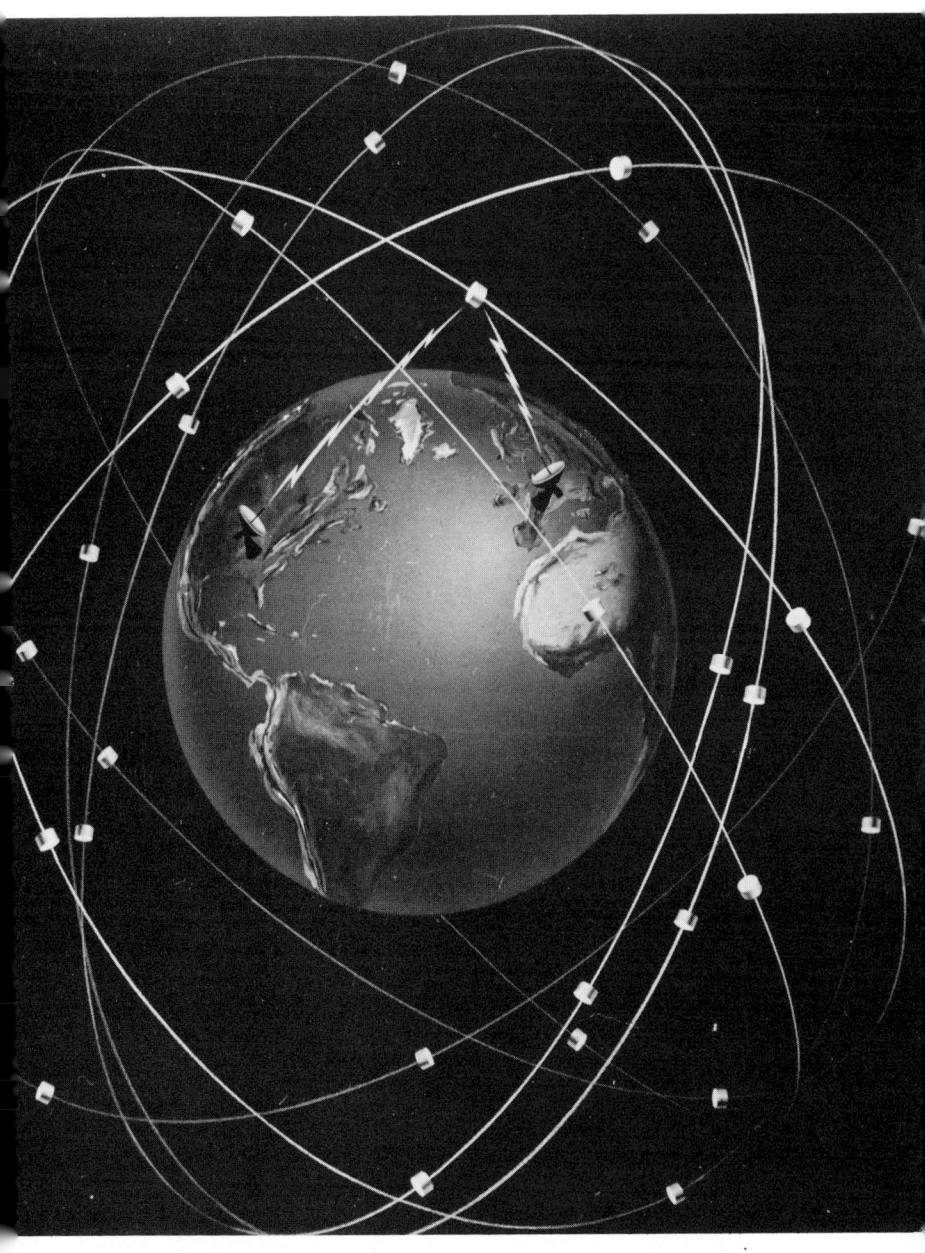

FIG. 7—A low-altitude active repeater communication satellite system is a possible alternative to the system shown in FIG. 6. (NASA photo.)

distribution, their energy levels, and the effect on various types of solid state devices must be measured in the space environment itself.

Other problems which need solution are concerned with the acceleration and vibration associated with launching, temperature control in space where energy can only be dissipated by radiation, and leakage or outgassing of components in the hard vacuum of space. There are many components which must be developed or improved to provide the long life and reliability necessary to a viable commercial satellite system. Batteries must be improved, and traveling wave tubes for power output stages must be developed. Solid state devices do not exist which can provide sufficient power at the required frequencies (2,000 to 8,000 megacycles). In short, there is a great deal to be done and much is under way. The currently planned flight program includes:

Echo II: A more rigid sphere than Echo I is being developed. With a diameter of 135 feet, it is made of a laminate of 0.0002-inch thick aluminum on either side of 0.00035-inch thick mylar. Echo II will be placed in a 650-mile circular polar orbit.

Rebound: A project to develop multiple satellite launching and placement techniques. Will place three Echo II spheres in a 1,500-mile circular polar orbit.

Relay: An active repeater satellite capable of wide band communications. Will be spin-stabilized and placed in an elliptical orbit 900 to 3,000 miles high. Relay will carry two complete transponders and a radiation measurement and solar cell damage experiment.

Telstar: An active repeater satellite capable of wide band communications. This is the American Telephone and Telegraph Company satellite. Will be spin-stabilized in a 500-mile to 3,000-mile elliptical orbit. Will carry a single transponder and a radiation measurement and solar cell damage experiment.

Syncom: A light-weight active repeater to be placed in a 30° inclined 24-hour orbit (altitude 22,300 miles). Spin-stabilized with spin axis perpendicular to the orbital plane, it will provide for period control and narrow band communications via its two transponders.

West Ford: An attempt to establish a rather sparse belt of orbiting dipoles in polar orbit.

The first steps have been taken. The next year will bring about a number of dramatic demonstrations and significant experiences, but it will be necessary to move cautiously. The cost of satellite systems is large, and a great deal of international planning must accompany the technical determination of how best to invest in this new promise, a promise of a truly global communication system capable of serving all mankind.

HAZARDS OF COMMUNICATIONS SATELLITES

JOHN R. PIERCE

To many people, communications satellites seem as new in concept as they are in physical fact, although the idea of communications satellites is over fifteen years old.

The idea of earth satellites for communication is credited to Arthur C. Clarke, who said in 1945:

> Although it is possible, by a suitable choice of frequencies and routes, to provide telephony circuits between any two points or regions of the earth for a large part of the time, long-distance communication is greatly hampered by the peculiarities of the ionosphere, and there are even occasions when it may be impossible.
>
> Unsatisfactory though the telephony and telegraph position is, that of television is far worse, since ionospheric transmission cannot be employed at all.
>
> It will be possible in a few more years to build radio controlled rockets which can be steered into such orbits beyond the limits of the atmosphere and left to broadcast scientific information back to the earth. A little later, manned rockets will be able to make similar flights with sufficient excess power to break the orbit and return to earth.
>
> It will be observed that one orbit, with a radius of 42,000 kilometers, has a period of exactly 24 hours. A body in such an orbit, if its plane coincided with that of the earth's equator, would revolve with the earth and would thus be stationary above the same spot on the planet. It would remain fixed in the sky of a whole hemisphere and unlike all other heavenly bodies would neither rise nor set. A body in a smaller orbit would revolve more quickly than the earth and so would rise in the west, as indeed happens with the inner moon of Mars.
>
> Using material ferried up by rockets, it would be possible to con-

Hazards of Communications Satellites

struct a "space station" in such an orbit. The station could be provided with living quarters, laboratories, and everything needed for the comfort of its crew, who would be relieved and provisioned by a regular rocket service.

A single station could only provide coverage to half the globe, and for a world service three would be required, though more could be readily utilized.

The power required for the broadcast service would thus be about 1.2 kilowatts.

When it is remembered that these figures relate to the broadcast service, the efficiency of the system will be realized. The point to point beam transmissions might need powers of only 10 watts or so.

It seems fairly certain that frequencies from, say, 50 to 100,000 megacycles per second could be used without undue absorption in the atmosphere or the ionosphere.

Naturally, there are ideas that can be questioned in Clarke's early, pioneering paper. Perhaps his estimates of the power required for broadcasting are somewhat low. In comparison with relaying messages across oceans, the use of broadcasting satellites seems rather difficult and not too attractive. Clarke thought in terms of 22,300-mile-high "stationary" satellites; today lower orbits seem to have some advantages. Clarke thought in terms of manned space stations; today these seem very remote, while experimental communications satellites have actually been realized. Clarke thought that frequencies up to 100,000 megacycles per second could be used; we now know that not only oxygen and water vapor absorption, but rain attenuation as well, make such frequencies unattractive.

THE AUTHOR'S VIEW IN 1955

In 1955 I discussed possibilities of satellite communication in a paper written before I had read Clarke's:

Two different sorts of satellite radio repeaters suggest themselves. One consists of enough spheres in relatively near orbits so that one of them is always in sight at the transmitting and receiving locations. This sphere isotropically scatters the transmitted signal, so one has merely to point the transmitter and receiver antennas at it to complete the path. Another system uses a plane mirror or an active repeater with a 24-hour orbital period, located directly above the equator at a radius of around 26,000 miles or an altitude of about 22,000 miles.

For instance, only transoceanic communication has been mentioned, and for a reason. There are at present transcontinental television circuits. The announced cost of the American Telephone and Telegraph Company's transcontinental TD2 microwave system was 40 million dollars. This is only 5 million dollars more than the 35 million for the 36-channel transatlantic cable, and yet the TD2 system provides a number of television channels in both directions, as well as many telephone channels. Perhaps even more important in an overland system, it provides facilities for dropping and adding channels along the route. Without such flexibility, an overland system would be almost useless.

The great advantages of the passive repeaters over active repeaters are potential channel capacity and flexibility. Once in place, passive repeaters could be used to provide an almost unlimited number of two-way channels between various points at various wavelengths. They would also allow for modifications and improvements in the ground equipment without changes in the repeater.

Spheres, which reflect isotropically, are the most flexible of passive repeaters, because they allow transmission between any two points in sight of them. Moreover, with spheres there is no problem of the angular orientation of the repeaters.

The disadvantage of passive repeaters is the great path loss, so that even assuming antennas of a difficult if not prohibitive diameter and accuracy, the power required is large, although probably attainable.

The attractive feature of an active repeater is the small power required and the small antenna needed at the repeater, as well as the small power required at the ground.

In conclusion, it can be said that, disregarding the feasibility of constructing and placing satellites, it seems that it might be possible to achieve broad-band transoceanic communication using satellite repeaters with any one of three general types of repeater: spheres at low altitudes, or a plane reflector or an active repeater in a 24-hour orbit (at an altitude of around 22,000 miles).

The proposal is for relaying across oceans rather than either a universal transmission service or broadcasting. Further, unlike Clarke's, the satellites are to carry no men, and perhaps not even radio equipment. In writing the paper, I did not appreciate how small an orienting effect the gravitational gradient has, in comparison with light pressure, for instance.

RECENT PROPOSALS ARE MORE MODERATE

In September, 1960, W. E. Morrow, Jr. wrote:

> This paper discusses the properties of a new type of long range radio communication technique which makes use of the scattering of microwave energy from a large number of scatterers distributed in orbit about the earth. This technique has the advantage that only two orbiting belts of scatterers are required to achieve global long range communication coverage. There are additional advantages of very modest ground antenna tracking requirements, high reliability, and the ability to support a very large number of independent circuits.

Here a satellite communication system is reduced to the simplest possible form—a swarm of dipoles rotating as a belt around the earth.

It is scarcely necessary to point out that the later proposals are less ambitious than those made earlier. This is not characteristic of all writings about satellite communication. Such conservatism reflects in part an increased appreciation of the uncertainties of space technology. It reflects in part the desire to achieve practical results quickly. The need for more transoceanic circuits is real and present.

In the first year of its service, the quality and magnitude of communication provided by the first transoceanic telephone cable of 1956 led to a 70 per cent increase in the number of transatlantic calls, and to longer calls as well. Last year overseas calls increased nearly 20 per cent. Cables have been laid to Hawaii, Alaska, Puerto Rico, and France; an additional cable of increased capacity to Europe is planned, as well as one to Japan. Overseas telephone service is expanding so rapidly that it alone could justify a satellite communication system.

Some argue that the large time difference between Europe and America might preclude transoceanic television. I do not believe that we will know until we actually have broad-band transoceanic circuits. Certainly, however, there will be ample uses for broad-band circuits for voice, data, and picture transmission. Transoceanic television will be tried and may succeed.

In 1945 and 1955 we had neither the urgent need for nor the ability to orbit communications satellites. Today Score, Echo, and Courier have been orbited. But we know that half of the launchings in our space program fail. Further, the life of many payloads, including Score and Courier, has been far short of that necessary in a useful communication, and Echo is an experiment, not an attempt to provide practical, economical communication. Much research and development lie between us and the realization of even the simplest practical satellite communications systems.

Satellite communication is an outgrowth of two arts. On the one hand, it is made possible by a new, shaky and very expensive art of space flight. This essential art has been brought into being at government expense through our missile and space programs, with other ends than satellite communication in view. A sensible satellite communication program must make use of existing vehicles. Even modifications of existing vehicles are technically hazardous as well as costly. The other art is a mature and manageable art of ground electronics and communications. The two arts are blended in the electronic payload.

PROBLEMS AND MEANS

Electronics in space suffers from a number of hazards. All equipment must withstand tens of g's of acceleration during launch. Ultimately, only radiation cooling is available; heat must be conducted to radiating surfaces. There can be no convection even in a pressurized container, for there is no gravity. In an alternate environment of sunlight and shadow, temperature variations are large.

Vacuum and gas at atmospheric pressure are good insulators, but pressures are encountered during ascent at which a discharge can be initiated at a comparatively low voltage. Once in orbit, pressurized containers may leak, and vapors from apparatus can produce appreciable gas densities. The seals on storage batteries sometimes leak.

Finally, radiation progressively reduces the efficiency of the solar cells on which space payloads depend for power so that transparent protective covers must be provided, and some degradation in efficiency allowed for. And our knowledge concerning the integrated amount of radiation a satellite encounters may be uncertain by an order of magnitude.

Undoubtedly, long life can be achieved in space, but to do so, we must go beyond the standards of radio and television sets and attain the reliability characteristic of the communications equipment of common carriers.

Beyond purely electronic hazards, some communications satellite proposals call for oriented satellites. It seems possible that very low satellites can be oriented passively by gravitational gradient, but how low is low, and when will this be demonstrated? Passive orientation by means of the earth's magnetic field may work up to the height at which the earth's magnetic field is stable, but not for 22,300-mile "stationary" satellites. The only method of orienting such satellites appears to be by the use of flywheels plus gas jets or other impulse-producing devices.

Such "active" orientation has been demonstrated for a few days, but not for the years called for by communications satellites. Much remains to be done. The operation of mechanisms in a vacuum is hazardous, and so are pressurized containers. If orientation is achieved or influenced by command, there is an added danger that the system will be activated by a foreign transmitter. If elaborate codes are used to avoid this, there is a great hazard that malfunction will make the equipment unresponsive to legitimate commands. These are not idle worries; space payloads and command systems have been sadly fallible in practice.

THE CASE FOR SIMPLICITY

It is clear that the hazards of satellite communication lie in the vehicle and its launching, and in the proper functioning of the payload. The expense of a satellite system will be largely the cost of launching and replacing satellites. Bugaboos about ground antennae and the tracking of non-stationary satellites have been raised, but I think that they are unwarranted. Echo was automatically tracked to within a tenth of a degree by means of orbital data. Because of its large area-to-mass ratio, Echo has had an unusually rapidly changing orbit, and undoubtedly with refined equipment a smaller satellite could be tracked considerably more accurately.

Under these circumstances, a strong case can be made for using a very simple satellite, even though this calls for more complicated ground equipment.

Should we go so far as to use passive balloon satellites like Echo, or a belt of dipoles? This seems to call for ground transmitter powers of several tens of kilowatts for television bandwidths. Such large powers increase the hazard of long-range interference with ground microwave systems. Dipoles form a multipath medium, and large bandwidths can be obtained only by tricky means. They are very attractive for providing minimum essential military communications, but less so for our already large civilian communications.

Light, low-power, active satellites are an attractive alternative. A light satellite means that one can ultimately launch more satellites per vehicle. As power governs weight, a low-power transmitter seems desirable. This has the further advantage that low-power microwave tubes last much longer than high-power tubes, and low-power satellite transmitters will not interfere with ground microwave systems, while high-power transmitters might do so.

ECONOMY IN SATELLITE PLANNING PROGRAMS

In order to get the most results with the least satellite equipment a sensitive receiver is required. In the Echo experiment, a receiving system noise temperature of about 24 degrees Kelvin is attained by using a ruby maser and horn reflector antenna, and this could be bettered. Below 1,000 megacycles per second, cosmic noise vitiates the advantages of such a receiver. Above 10,000 megacycles per second the absorption of atmospheric oxygen and water vapor lead to appreciable thermal noise from the atmosphere, and thermal noise during rain would cause increasing interruption of service with increasing frequency.

Besides sensitive receivers, one can use broad-band modulation schemes in reducing the power required aboard the satellite. Frequency modulation with an index of 5 gives a 14 decibel gain (25 times), but calls for a bandwidth of about 50 megacycles per second to transmit a television signal or 600 telephone channels. Although broad-band modulation reduces the number of separate frequency channels in a given block of frequencies, it also reduces interference between systems operating on the same frequency and would make it possible to use the same frequency channel for simultaneous transmission to, or reception from, satellites in many parts of the sky.

By using a sensitive maser receiver together with an antenna 75 feet in diameter, and a broad-band form of modulation, one could transmit a television signal from a 2,500-mile-high omnidirectional satellite antenna to ground with about a watt of power. The same power would suffice for a 24-hour "stationary" satellite providing that the satellite made use of a directional antenna with a beamwidth just broad enough to cover the earth, or for, say, a 7,000-mile-high satellite with some simple means of orientation.

One satellite system which has been proposed uses low altitude satellites (7,000 miles high), with passive or no orientation, in large enough numbers (some multiple of ten) so that by chance one satellite is almost always available for each path. Such a system is subject to predictable, infrequent outages. While many satellites are required, a given satellite could successively serve different paths in different parts of the world. The loss of a few satellites would degrade the performance of such a system inappreciably. Presumably, such satellites would be launched from time to time in batches of perhaps ten.

Another satellite system, like that in the Army's Advent program, makes use of a few 24-hour "stationary" satellites, complete with orientation, station-keeping, and command control. Proponents of such a system sometimes talk in terms of three such satellites. This would be hazardous to the point of being ridiculous, for the loss of a single satellite would disrupt some paths completely. Such a satellite system has a disadvantage for common carrier telephony. The round-trip delay of about half a second causes trouble with the echo suppressors that are used on all long circuits to interrupt incoming signals when the user is talking—a sort of trouble which users of transatlantic cables are spared.

FORECAST FOR SATELLITE COMMUNICATION

Very sketchily, then, such are the history and the technical prospects of satellite communication. What other problems does it face? What of its future?

It is clear after even the most superficial examination that satellite communication must first be used to link the extensive land communications which already exist. To speak of any initial application in serving underdeveloped areas would be like a proposal (made during World War II) that the atom bomb be built in some African or South American country because it would be good for that country's economy. Whatever benefits developing nations will gain from satellite communication must spread out to them from initial successful use by more highly developed economies and technologies. We must do first things first, if we are to succeed at all.

Let us turn to the problem of linking, say, Western Europe with America by means of a satellite communication system. One problem, of course, is that of frequencies.

The frequencies needed are the sort of microwave frequencies used for a wide variety of military and civilian purposes. To be useful, a satellite system must form a part of the complex existing network of common carrier communications. It seems clear that in order to avoid delays and intolerable conflicts, satellite communication must initially make use of frequencies already reserved for, and used by, common carrier companies and agencies. These companies and agencies have a good chance of avoiding interference with their communications, although this will mean not using certain frequencies for microwave radio relay within about 100 miles of every satellite ground terminal. Eventually, more frequencies will be needed. Effective transoceanic communication using satellites will always require that the use of microwave frequencies for communication over land somehow be limited and controlled. This is a difficult problem now faced by national and international organizations, among them the century-old International Telecommunications Union, now a part of the United Nations.

"OWNERSHIP" OF SATELLITES

Another and non-technological problem is that of ownership and operation of the satellites (if one *can* own a satellite) and of the ground equipment used in satellite communication. Traditionally, transoceanic communications have been carried out co-operatively on a non-political basis by the various organizations which provide telephone and telegraph service. In most countries these are departments of the government, usually the post office, as in Great Britain, France, and Germany. In some countries, as in Canada, the organization is a government-owned corporation. In our own country, regulated private corporations provide telephone and telegraph service.

In the case of transoceanic radio, in each country the organization which provides telephone or telegraph service owns and operates the shortwave transmitters and receivers in its territory, whether this be the Soviet Union, Great Britain, or the United States. In the case of telephone cables, the cable is paid for, and owned jointly by, the telephone organizations of the countries which it links. This seemingly complicated ownership of international communication facilities has a long tradition of successful operation among countries of the most widely varying and antagonistic politics. In some countries it is a matter of law as well as of tradition. It has been little affected even by the cold war.

There is no financial or technological obstacle to following the same pattern in satellite communication. The initial cost of a satellite communication system will be comparable to that of a large transoceanic cable system. The American Telephone and

Telegraph Company (AT&T) has announced that it is prepared to bear the costs of communications satellite research, including launchings, and its share, together with foreign partners, of the cost of an international satellite communications system. An agreement has been reached with National Aeronautics and Space Administration (NASA) for launching experimental satellites (at AT&T expense) in 1962, and the Bell Telephone Laboratories are vigorously pursuing a program of satellite research and development. Technologically, satellites are merely a new and potentially powerful way of enlarging the world's means of international communication.

The military are, of course, interested in satellite communication. Besides the Army's Advent program, the Air Force plans an experimental launching of dipole reflectors in its Project West Ford. Further, NASA is sponsoring satellite research. Its Project Relay calls for the launching of an experimental satellite for transoceanic television and telephony in 1962, and NASA plans to support other satellite projects as well.

THE POLITICAL PROBLEM

If the research now under way can be pursued without hindrance, and if communications satellites can somehow be integrated into the world's present communication facilities, we can look forward to many benefits. Unhappily, there are people who regard the science and technology of space as not just one aspect of the world's long course of scientific and technological progress, but as a unique political battleground where blows may be struck against foreign rivals or native institutions, where enemies may be intimidated and constituents cajoled. Should we use this strange and unique resource of satellite communication against unfriendly countries, or give it away to make friends? Should we not use the special science of space to supplant existing scientific and technological resources, and start fresh in a new and attractive pattern of national and international communication?

Such an attitude is dangerous. In divorcing space science and technology from the centuries of science and technology of which it is a part, we not only misinterpret its nature; we also abandon successful ways of dealing with real national and international problems and raise a host of new problems which are not truly relevant to the exploitation of space.

Satellite communication has already added to our prestige. But, satellite communication is no mere trick or weapon of propaganda. It promises substantial benefits to the citizens of all nations. It promises prompt, adequate, and economical transoceanic telephony. Beyond this, it promises the immediacy of television for many places not now reached by television.

Achieving these things is not going to be easy. Overambitious and technically marginal plans could block progress. Extraneous issues are perhaps even more hazardous. It will be hard enough to work out a scheme for international satellite communication without our insisting that the scheme must aid backward nations immediately, put the United States or the United Nations in the communication business, or serve as a political issue.

Happily, on July 24, 1961, President Kennedy announced a national policy on communications satellites which is consistent with their effective and rapid development and use in improving and extending international communication. Mention of the International Telecommunications Union in the statement is a wise acknowledgment of effective existing means of international co-operation. Despite the fact that President Kennedy's policy has been strongly attacked, it may be hoped that the benefits of satellite communication will be made available to the people of the world, perhaps not as soon as we might wish, but as quickly as scientific and technical problems allow.

SUGGESTED READING

CLARKE, A. C. "Extraterrestrial Relays," *Wireless World,* 51 (October, 1945), 305–8.

Communication Satellites. Hearings before the Committee on Science and Astronautics of the U.S. House of Representatives, 87th Cong., 1st sess., No. 19, Part 1, May 8–10, and July 13, 1961; Part 2, July 14, 17, and August 1, 9–10, 1961. Washington: U.S. Government Printing Office.

ORDWAY, F. I., III (ed). *Advances in Space Science.* Vol. I. New York: Academic Press, 1959.

PIERCE, J. R. "Communication Satellites," *Scientific American,* 205 (October, 1961), 90–102.

———. "Orbital Radio Relays," *Jet Propulsion,* 25 (April, 1955), 153–57.

"Project Echo," *Bell System Technical Journal,* 40 (July, 1961).

Satellites for World Communication. Hearings before the Committee on Science and Astronautics of the U.S. House of Representatives, 86th Cong., 1st sess., No. 9, March 3–4, 1959. Washington: U.S. Government Printing Office.

PART TWO

Space Research

Introduction
Biology and the Space Environment
Challenges to Biology
Flying Telescopes
Rocket Probes
The Earth and Near Space
The Sun
The Moon and Planets

INTRODUCTION

JOHN A. SIMPSON

Technological development of artificial satellites and deep space probes has already convinced us that they are of practical importance for both commercial and military applications. In commerce they will shortly provide unprecedented means for communicating continuously over the world, handling sufficient volumes of information to meet the demands of the population explosion. For national defense, they change the policy of *open skies* to a policy of an *open planet;* they have already yielded new concepts for practical weather predictions, crucial for military decisions and the agricultural economy of our nation. Space vehicles are also a real, though unevaluated, prestige and political factor in the struggle between the East and the West for the minds of men in the neutral and the new nations. These developments have come about without placing a strain on our present storehouse of scientific knowledge.

Though this technology will become a vital factor in our economy and for our immediate national needs, the most important consequences of these developments are for the exploration of the inner solar system, and for fundamental scientific discoveries in the physical and biological sciences. There is a dichotomy of purpose between exploration and scientific investigation, yet these two objectives support each other and are being simultaneously pursued with great benefit. For the scientific investigator, a satellite or deep space probe system becomes a practical laboratory tool to assist him in investigations—just as the microscope, cyclotron, or telescope are tools for our sciences at present.

In the long run of history great advances for man will be found in exploration and scientific discovery in space, resulting in radically new ideas about our solar system, galaxy, and the universe. Although handicapped by small payload capacity, the United States space probes and satellites have yielded an astonishing variety of scientific discoveries important not only to the advancement of space science, but to the whole body of science and technology. On the other hand, the dramatic lead of the Soviet Union, demonstrated through the imaginative application of its advanced space technology and engineering, has proved to the world that man can reach the moon and the planets — via instruments alone or by going there himself. We may evaluate the present relative positions of the two nations as follows: For the years of the 1960's, the greatest scientific advances will be made with instrumented satellites and space probes. The United States can continue its lead in scientific discoveries for man if it is determined to do so. A man in orbit is not needed to control the instruments envisaged for the next six to eight years; in fact, man in space becomes a serious liability in view of his gross motions and the extra instrumentation required to keep him alive. However, in the long run, for direct exploration at great distances, such as of Mars, the assistance of man on the site will be required.

VAN ALLEN RADIATION BELTS

The first important discovery using an artificial satellite was made by James Van Allen at the State University of Iowa, who detected fast-moving protons and electrons trapped in the geomagnetic field.

We know that the magnetic field of a bar magnet, or the earth, may be represented by drawing lines of force extending from one pole of the magnet to the other. For the earth, a model of such series of lines of force is represented by sprinkling iron filings on a sheet of paper held over a bar magnet so that the iron filings line up along the directions of the line of force. We also know that if a wire is carrying an electric current and this wire is suitably placed in the magnetic field, a force will act upon the wire because of the motion of the electronic charges in the metal wire. These same forces come into play when free electrons, or protons (the nuclei of hydrogen atoms), move with high velocity in the external magnetic field of the earth. The

Introduction 77

forces act on the charged particles in such a way that the particles spiral about the lines of force, with the spiral radius diminishing as the magnetic field increases or as the lines of force come closer together. The particles are trapped in the magnetic field to follow a spiral path from one hemisphere to the other for long periods of time. Therefore, it does not take many particles to build up an appreciable stream of electrons confined to the lines of force. Since these particles also drift around the earth, they form shells, or belts—the Van Allen belts. At first it was thought that the great outer belt was of solar origin, but recent experiments show it likely that the particles gain most or all of their energy while they are trapped on the lines of force of the earth's field. Close to the earth there is a smaller belt which is dominated by circulating protons. A study of these trapped particles will undoubtedly yield information on how they gain their energy and, in turn, increase our knowledge of magnetic field disturbances which have important consequences for us on the earth.

COSMIC RAYS AND "WIND" IN SPACE

For fifty years we have known that very high energy particles come directly to us from far beyond the influence of the earth and its magnetic field. These particles have very high energies and consequently travel with nearly the velocity of light. Called cosmic rays, they consist mostly of the nuclei of hydrogen atoms (protons), although about 10 per cent of the particles are the nuclei of helium (alpha particles). In addition, there is a small contribution of nuclei of the elements found in nature, at least up to iron in atomic weight. How these particles gain their energy and where they achieve their energy gain constitutes one of the fundamental problems in physics today. For many years they have been investigated at the surface of the earth, on mountain tops, with jet aircraft, balloons, and small rockets. The introduction of the space vehicle, which can completely escape the environment of the earth, unshackles the scientific investigator and enables him for the first time to perform direct experiments in space concerned with the origin of these particles. It appears that those of the highest energies come almost exclusively from outside the solar system. Mixed in with these particles are occasional bursts of energetic protons and alpha particles which come from solar explosions on the sun. It has been shown that the particles from the sun have different energy distributions and other properties which distinguish them from the galactic cosmic rays. At the time of this writing, the only known and uncontrolled hazard to man in interplanetary space is the proton beam which comes from these solar flare explosions. Therefore, the hazards for an unprotected man in space are directly related to the phase of the solar activity cycle.

ONE MYSTERY, TWO HYPOTHESES

One of the puzzles in cosmic ray physics was the observation that although most of the radiation was coming from outside the solar system, and hence was constant over long periods of time, the intensity actually varied by a large factor throughout the time of the solar cycle. Experiments on the earth convinced investigators that they were here dealing with a phenomenon in which the sun controlled the galactic cosmic ray intensity. One possibility was that the sun provided screening magnetic fields of some kind which prevented the particles from reaching the vicinity of the earth. Two extreme models were used to test two possibilities. The screening might be around the earth, while the full intensity of cosmic radiation existed elsewhere in the solar system. It was also possible that the magnetic screening occurred in such a way that the intensity throughout the entire solar system was depressed during some periods of the solar cycle. Direct experimentation with deep space probes proved that the latter alternative is close to the physical situation.

To perform this experiment, it was necessary to measure the galactic cosmic ray intensity from the vicinity of the earth out to great distances in space at a time when the intensity at the earth was known to be low. The question to be asked on the space probe was: Does the intensity increase as the detectors are carried away from the earth until they reach the full level of galactic intensity at great distances in interplanetary

Introduction 79

space, or does the intensity remain low even though the instruments are carried far beyond the influence of the earth?

ANSWERS FROM PIONEER V

The answer was obtained by the United States deep space probe Pioneer V, launched in March, 1960. The space probe succeeded in providing data for scientific investigators for up to 10 per cent of the distance from the earth to the sun. The technological achievements were as remarkable as the scientific results. At the time of this writing, this is still the farthest distance over which scientific experiments, and the telemetry of the data from experiments, has been carried out by man. Pioneer V weighed about 95 pounds and of this weight only about 10.5 pounds were available for the direct scientific experiments.

The probe revealed that the galactic cosmic ray intensity decrease was not only in the vicinity of the earth but also throughout the entire inner solar system. Let us consider an analogous model—a large tub of water where the water molecules represent cosmic ray particles and the height of the water surface is a measure of cosmic ray intensity. Some point on the surface represents the location of the sun and the inner solar system. The sun's heat causes speeding charged particles to keep escaping from the solar atmosphere as a "wind" to penetrate interplanetary space. This solar wind moving out from the sun can be represented by a hose through which an air stream escapes. Approaching the surface of the water from above with the air stream one would see a dimple formed as the water is pushed radially away from the axis of the air stream. At the center of the air blast the water is at a much lower level than at great distances away. By varying the velocity and quantity of air flow, it is possible to simulate variations in the eleven-year cycle of solar activity in which the solar wind changes. During minimum solar activity, the dimple in the water surface will practically disappear and the water level will appear at its full height everywhere. Similarly, galactic cosmic rays at the minimum of solar activity will penetrate with full intensity to the vicinity of the earth and the inner solar system.

SOLAR WIND AND SCALE SIZE IN NATURE

In the case of the solar wind in interplanetary space, it is not the charged particles of the wind which interact with the cosmic rays, since they do not collide with each other. Instead, because of the heat of gas escaping from the sun, all the particles remain charged, and moving charges may produce magnetic fields. A great deal of work has been done in the past decade to prove that moving streams of ionized gases—plasmas—may transport with them magnetic fields. These magnetic fields, either associated with the sun or with the plasmas themselves, interact with the oncoming galactic cosmic ray charged particles, deflecting them in such a way that there is a net decrease of intensity close to the sun at maximum activity.

These experiments in space emphasize another fact, namely, that there occur phenomena in nature which cannot be scaled down—miniaturized—without losing the essence of their physical properties. Within the laboratory no arbitrarily small amount of uranium 235 can be studied which will provide the direct proof that a nuclear reactor will "go." The scale of experiments must be sufficiently large or the observations cannot be made. In space sciences, this scale in many cases is the size of the entire inner solar system. In other cases, it is only the scale size of the earth and its magnetic field.

Without doing justice to the many other exciting discoveries which have been made in space experiments, these few have been selected as an introduction into a general area of physics now opening through the use of space vehicles, an area vital to the understanding of astrophysical phenomena. The artificial satellite and deep space probe systems have become tools to extend man's concepts of the physical universe and his place in it.

BIOLOGY AND THE SPACE ENVIRONMENT

COLIN S. PITTENDRIGH

Progress in science depends in part on our ability to enrich and extend the limited array of sense organs that natural selection has given us in the history of life. We have had to build special transducers to "see" any but that tiny fraction of the electromagnetic spectrum we call "visible" light; we must supplement the lens system of the eye to see all we wish to see of stars or in living cells. And to explore the atomic nucleus we need the whole elaborate technology of accelerators, bubble chambers, emulsions, and the like to channel information into our limited sensory modalities.

The scientist, aware of this, sees the significance of the new space technology very differently from the layman. For the latter, big rockets and manning space vehicles fall in the same category as climbing mountains—space is a super-Everest; it is there and if we can, we should conquer it. The scientist may well sense this challenge also, but basically for him the rockets, combined with automatic control systems and telemetry, are nothing more than a new and spectacular extension of his sensory capacities. In geophysics, meteorology, and astronomy the investigator now has access to new kinds of observation and information from which his earthbound existence previously barred him. It is certain that in all these areas of study we are on the threshold of major discovery; and in some areas, like that of fields and particles, we are already over the threshold. Some physicists may regard "major discovery" too weak a phrase; in some cases "revolution" would be a better word. This view is based on the belief, or at least the hope, that we shall gain more than just new information on known features of the physical world; we are, it is felt, in for some real surprises. At any rate the latter prospect is not excluded, for there may well be aspects of the physical world which we can detect only in space by means of our new sensory probes.

What is the prospect for a similar impact of space technology on the life sciences? This is not an easy question to answer. On the one hand, the biologist might insist that only with the space technology we hope to have in the next ten years does the most exciting scientific question of the day—the existence of life on other heavenly bodies—enter the realm of the real, the answerable. For this reason alone he might insist that the importance of the new technology is as great for biology as for physics. On the other hand, he could pass this off as at best a very long shot and insist that his real business concerns a particular class of complex systems—living organisms—that have evolved and can function only *here*, on this planet; space holds no interest for him, as biologist.

A proper perspective falls somewhere between these extreme views. But let us admit right away that space technology holds far less promise of any revolution in biology than in physics. And even "major discovery" (the long shots about extraterrestrial life aside) is more than most biologists (the writer included) would be willing to allow as a sober guess. It is one of the delights of science, however, that sober guesses are not always correct!

The following chapter on exobiology by Novick and Lederberg therefore deprives me of an excuse to discuss the only foreseeable, first-rate question in biology for which we need spacecraft. There are, however, other less spectacular ways in which space technology will have an impact on the life sciences, and it is my purpose to outline these briefly.

MAN IN SPACE

The most obvious influence derives from the man-in-space projects, Mercury and Apollo, and especially from the latter, which calls for a vehicle to house several men in a simulated terrestrial environment adequate for a prolonged journey—such as to Mars and back. Both projects are controversial; are they worthwhile? What is the scientific, as against political, justification for what some feel is essentially a circus stunt? The critic can point out that now and for years to come we can do an enormous amount of useful work in the near reaches of space with unmanned probes. Separated from our new sense organs by what seem—today—immense distances, we can, by radio communication, remain part of the control loop regulating their performance. But the advocates of Apollo argue that there is a practical limit to the distances we may explore in this way; and this limit hinges on the fact that the velocity of light is finite.

The precision of response of our probes will demand a "reaction time" that cannot be realized so long as we continue to remain part of their control loop at those immense distances involved in studying Mars and the more distant planets. At these distances it will take times of the order of several minutes to an hour for information to complete the vehicle-to-man (on earth)-to-vehicle circuit; and such time lags, it is argued, preclude the necessary precision of response. Apollo, then, is not a circus stunt; it is a technological necessity if we are to explore the remoter regions of space. Man must go *with* his new sense organs.

This is the strongest argument we can give for the man-in-space program, and even so it is obviously not a scientific (biological) goal in itself. It is a formidable piece of biological engineering to be achieved as a necessary part of the new technology of space probes. But as such it is bound to have a considerable influence, direct and indirect, on physiology.

Man is a terrestrial organism; he functions normally only in a particular environment—that of earth—and to enter space he must carry that environment with him. The plain fact is that the biologist is not yet in a position to inform the engineer fully which features of our environment he must simulate, and how precisely; environmental, or ecological, physiology is not that well developed. There is a simple reason for this. The physiologist has found it convenient to stabilize the more obvious, powerful environmental inputs and then, ignoring them as a constant, to proceed with analyzing the organism's metabolic and genetic machinery. For the most part he has gained insight into the role of environmental factors only as he exploited them as perturbations of the organism which has been his real object of interest. Thus while we can now specify optimal conditions of first order variables like temperature and oxygen and carbon dioxide tensions, we have difficulty being precise about much else. It is not that we do not realize there are other variables to consider; it is simply that in the growth of physiology so far they have received the neglect that is proper for second order effects.

ENVIRONMENTAL PHYSIOLOGY

There are physiological effects on multicellular organisms due to barometric pressure, but they have been very poorly studied. There are recurrent claims that the degree of ionization of atmospheric gases is also important, but I know of no physiologist who would commit himself on the point. Do we need, in our simulation of earth's environment, to provide the heavy inert gases of the atmosphere? Possibly not, but we do not know with certainty, since the work necessary to answer the question has not been done. Similar ignorance leaves a host of other questions unanswerable: What is the effect of magnetic fields, of radio frequency fields, of low voltage direct current fields, of coriolis forces, and the like? What environmental inputs, what sensory modalities are involved in the bicoordinate navigation of birds?

All these questions have attracted good biologists in their time, but all remain embarrassing to the environmental physiologist. The possible role of magnetic fields and coriolis forces were both investigated seriously in the postwar years that saw a great upsurge of interest in the long obscure problem of bird navigation. Both have now been dropped again—largely because most of the workers became distracted by the spectacular discovery that birds use the sun azimuth as a compass and possess the internal chronometer necessary to compute its hourly displacement. But the whole field of animal navigation still involves many mysteries. Some birds can perform true bicoordinate navigation; discovery of clock and compass is therefore not the full solution. They have access to another coordinate system still unidentified. Precht, at Kiel University in Germany, has been reporting that seagulls displaced by many miles will show a true heading for

Biology and the Space Environment 85

home even within a tent that excludes all visual celestial clues. It may be, of course, that all this remarkable behavior will ultimately yield to analysis in terms of known sensory modalities; but it would be foolish to assert this will certainly be so.

Of course even this brief list of poorly known, or completely unidentified, environmental inputs and sensory modalities surely includes some we shall ultimately be able to ignore in simulating a normal environment. But just as surely we cannot afford to accept present ignorance on them all; some will prove significant even in the real terrestrial environment and may be more so in the necessarily incomplete and distorted simulacrum we can achieve in a space capsule. This is probably the most obvious and certain impact on biology we can expect from the space program; it imposes on the physiologist the responsibility to explore fully those little known areas of sensory and environmental physiology he has been able previously to bypass.

One reason why environmental or ecological physiology has not attracted more attention is the sheer diversity of actual micro-environments exploited by this planet's organisms; they range from the pages of library books, through the ear cavities of moths, to the water of hot springs. A full analysis of this known diversity is out of the question, and in attempting to focus on truly general features of living function, the main tradition in physiology has tended to neglect environmental considerations as though they were all second order complications. But there *are* really *general* aspects of the environment on this planet, and they have tended to suffer the same neglect. One dividend we can hope for in the space program is that general environmental aspects will now receive their due attention.

IMPORTANT EXAMPLES

Two examples will suffice to illustrate these general and poorly known aspects of environmental physiology. One relates to the effects of the earth's rotation and the other to its gravitational field. The first concerns what used to be called "persistent daily rhythms." The fact that organisms characteristically do some things and not others at particular times of day is both too general and too obvious to excite much interest on the part of the physiologist. The assumption is easy and reasonable that the change in activity is causally determined by changing environmental inputs. It has been known for over 200 years that this is not the whole story. The French astronomer De Mairan observed in 1729 that the daily up and down movements of plant leaves persisted in an environment with essentially no periodicity of light and temperature.

The subsequent history of this phenomenon of persistent daily oscillations in function is too complex to review even briefly here. It has included discussion of whether the oscillation was being forced by unknown geophysical periodicities, but by now all but one of the many laboratories devoted to the problem have rejected this view; the persistent oscillations have a frequency that differs significantly—by minutes or hours—from that of the earth's rotation. The tendency has been to regard these *circadian* (from the Latin *circa diem*; period about that of a day) oscillations of leaf movement, locomotion, or sleep-wakefulness as at best adaptive curiosities superficial to the main physiological architecture of the organism which has been treated as aperiodic. Work in many laboratories over the last ten years is giving the lie to this view. The evidence has accumulated now to the point of certainty that these circadian oscillations pervade entire physiological and biochemical systems which act as innate oscillators whose frequency has evolved to match that of the physical environment. The environmental inputs that regulate this oscillator—the light and temperature cycles—are only acting as entraining periodicities regulating phase and period of the living oscillator; they are not forcing the vibration.

PROBLEMS FOR THE ENGINEER

My point is that this is a highly significant parameter of physiological organization; interference with these oscillations commonly imposes physiological damage—a fact which the engineer of man in space should know. Historical neglect of the phenomenon has in part stemmed from a neglect of that fundamental premise of biology that living systems are highly organized with respect to their environment. We may, as general physiologists, temporize in explaining the special environmental physiology of book lice but, in so shelving ecological questions as special, we are in danger of missing major effects due to truly general characteristics of the environment like its 24-hour periodicity—and perhaps its gravitational features.

The possible role of the earth's gravitational field in normal physiological function is my second, wholly problematic, example. Some organisms—in caves and the deep sea—escape almost entirely the influences of the earth's rotation, and we have still too little information to know whether they have been able to shed the circadian oscillations that characterized the physiology of their ancestors, who were fully exposed to detectable effects of the earth's rotation. But there is presumably no organism on this planet that escapes a gravitational effect of about one gravity. Plants and animals sense the direction of the gravitational field, and their behavioral responses to this input have long been studied by plant and animal physiologists. But neither of the latter have any information on the possible extent to which the general functioning (vis-a-vis special behavioral response) of the system depends on the existence of the gravitational field. Here, then, is a clear-cut case where we need the missing information for the engineering goal of man in space; where the information once obtained may well be of considerable theoretical significance, and where the information cannot be obtained without the existence of space vehicles to carry organisms out of the earth's gravitational field.

SURPRISES EXPECTED

The preceding emphasizes what I believe the space program's impact will be, or should be, on the progress of ground-based physiology. This emphasis is deliberate, because the general tendency is to assume that the availability of the new space tools will be the principal source, as in physics, of the anticipated new information. While I doubt that this is the case, except for exobiology, we must nevertheless not underestimate the role the vehicles can play. Until we know much more, we are not in a position to use space vehicles for crucial experiments that answer clear-cut questions. But there is more to the real business of science than ideally clear questions and ideally crucial observations. There is the role of plain exploration; let us see what happens. And it is a fact that we already have one biological experiment that has surprised us: Mutation rates in *Neurospora* placed in the Van Allen belt were found to be orders of magnitude higher than one could reasonably anticipate from the simultaneously observed radiation. This experiment was far from satisfactory as its author, De Busk, himself emphasizes, but it is already a challenge and a stimulus. It is a reminder that we can yet be really surprised by the biology of the space environment and for that reason we should temper our conservatism. If the physicist, armed with an incomparably tighter theoretical scheme than we possess, is ready for surprises we would be presumptuous, to say the least, to spurn such an attitude.

CHALLENGES TO BIOLOGY

AARON NOVICK and JOSHUA LEDERBERG

Biology, in contrast to physics and chemistry, has until now concerned itself with the phenomena on earth. But with the possibility of examining the nearby planets at first hand the biologist will soon face a wider perspective. He will have an opportunity to seek answers to the intriguing question of whether extraterrestrial life exists, and from studies of other planets he may gain information on the origin of life on earth. Moreover, the biologist will have a new and serious responsibility, not just to science but to the world as a whole, arising from the possibility that there may be biological consequences of space exploration itself.

In order to ask whether life occurs on another planet we must provide ourselves with a definition of "life." The consensus so far, derived from our knowledge of terrestrial life, is that any definition of life must be arbitrary; that is, if life has gradually evolved from inanimate matter, the demarcation between nonliving and living is a matter of judgment. Many biologists seek a convenient working definition of life in the fact that life exhibits the unique capacity to become more complex with time, that is, to evolve. A system to be described as alive must be able not only to reproduce itself but also to mutate, that is, to undergo randomly introduced alterations which are then reproduced. In any particular environment, individuals having certain combinations of characteristics will reproduce more rapidly than others and thus predominate. In time the species will have altered, adapting to the environment—which is evolution. The complexity that individuals can achieve is determined by the range of possible mutations, and if the range itself can be increased by mutation there are no foreseeable limits to the complexity that can be reached.

In the last few decades we have learned that all life on earth has fundamentally the same chemical basis. All forms of earth life depend upon an aqueous environment, and on moderate temperatures that permit organic macromolecules to be reasonably stable. Furthermore, they are all composed of similar molecules. We now know that two of these, the nucleic acids and the proteins, account for the most basic activities we associate with life. The proteins serve as specific catalysts, directing cellular chemical reactions that produce the chemical substances needed for growth. Proteins are constructed of relatively small subunits called amino acids, of which there are twenty different kinds, arranged somewhat like links in a chain. A typical protein may be composed of several hundred amino acid units ordered in a sequence specific for this protein.

Any organism must be able to make thousands of different protein molecules, and it derives information for their structures from its genetic endowment. It is now known that deoxyribonucleic acid (called DNA) forms the essential chemical basis of this endowment. DNA has the remarkable property of being able to contain information and to duplicate itself. From its structure, discovered by Watson and Crick, and from experimental studies, it is becoming apparent how DNA can carry out these functions. It is made up of four different subunits (nucleotides), with hundreds of nucleotides forming a long chain. It is believed that the arrangement of the four different types of nucleotides in a particular nucleic acid determines by some coding relationship the arrangement of amino acids in a particular protein. Furthermore, the DNA molecule is built of two complementary long nucleotide chains wound about each other, the sequence in one chain determining the sequence in the other. It is believed that each chain of a pair provides information for the formation of a new complementary chain; thus one pair of chains can make two pairs.

Mutation can occur if a wrong nucleotide is put in at some point. Such a change would lead to an alteration in the protein specified by this nucleic acid and would be perpetuated in the progeny. This relationship between nucleic acids and proteins forms the basis of all known terrestrial life, and the discovery of this relationship is a major triumph of biology. But much

Challenges to Biology

remains to be learned of the actual chemical mechanisms involved.

The origin of life has been the subject of much thought since the earliest speculations of man. A major early milestone in the development of biology was the demonstration that the suspected cases of spontaneous generation of life from inanimate matter were false and that recognizable life apparently always develops from an earlier parent. This led to the idea that prevailed for many years that life had its origin on earth in some extremely unlikely accident.

But more recent ideas have lent encouragement to the belief that there is a good, rather than unlikely, chance for life to develop on a planet like earth. It is argued that spontaneous chemical processes would lead to the formation of many complex molecules. In fact, amino acids are formed by the action of electric discharges on gas mixtures similar in composition to the presumed primitive atmosphere of earth. In the absence of voracious organisms such complex compounds would accumulate, especially in the oceans where they would be protected from the decomposing action of solar radiation. In this "soup" there could develop self-replicating structures that would catalyze their own formation. How this might have happened is not yet understood in any detail, but we are beginning to visualize the essential conditions for chemical replication. Much is being learned from biochemical studies of nucleic acids and from industrial syntheses of stereospecific polymers. It is possible that ancient rocks may yield "chemical fossils" from this early period in the evolution of life on earth which would help toward an understanding. Probably, these will be hard to find, since most such relics have been destroyed on the earth's surface by later occurring life, and beneath the surface by the action of high pressures and temperatures.

One theory of the origin of life, which seems less credible now, but which is difficult to rule out completely, is the hypothesis of panspermia advanced by the physical chemist Svante Arrhenius. He suggested that life originated on earth through the migration of spores to earth from some other planet. But it is difficult to account for the escape of a spore from a planet, and the idea that light pressure could give sufficient velocity to permit escape from the gravitational field of a planet as large as earth, or any planet large enough to sustain a significant atmosphere, is implausible except for objects the size of the smallest viruses. Moreover, assuming that a spore did escape from a planet, the intense solar radiation prevailing in the absence of a protecting atmosphere would destroy it in a tiny fraction of the time required to go from one planet to another. Finally, the hypothesis of panspermia only defers the problem of the origin of life to some unknown site. Nevertheless, considering future developments in rocketry, it would not be wise to rule out completely the possibility of an artificial panspermia where life itself evolves the means for making an interplanetary voyage.

From our present understanding of terrestrial life we can deduce some minimal conditions for life of any kind to exist. There must be a chemical system in an environment where large information-bearing molecules are stable; and in addition there must be a source of energy and a source of chemical raw materials in order that the information-bearing molecules can duplicate themselves. Further, the development of life would almost certainly depend on the presence of a liquid solvent like water in which chemical reactions would proceed far more rapidly than in the dry state.

EXTRATERRESTRIAL LIFE

On the basis of this picture of terrestrial life and of the mechanism of its origin, the biologist has many questions he would like to ask. Principally, of course, he wants to know whether life exists outside the earth. Should any be discovered, the biologist's first wish would be to gain an understanding of the chemical basis of this life. Does it, like earth life, use the protein-nucleic acid system? If not, what chemical systems are used for information storage and catalysis? Along with the search for life there is the additional important objective of finding, through studies of the chemistry of planetary surfaces, evidence that might help us understand better the chemical evolution which led to the appearance of life on earth.

Early missions will have severe limits on the weight of experimental apparatus that they can carry and on the amount of information that they can return to earth. It will have to be decided whether to gamble for a quick answer with a "long shot" experiment or use the early space probes for gathering more precise and certain information as a basis for the design of more sophisticated experiments. It does appear, however, that a popular choice for an early mission making a "soft landing" will be a visual examination of the surface of a planet. It is a thrilling prospect to examine close-up photographs of the surface of a planet like Mars, and such pictures should also serve as valuable guides for the design of later experiments. With great luck a close-up might give convincing evidence of Martian life. Along with chemical stains which give specific colors with substances like nucleic acid and protein, this photographic equipment could also be used to identify the chemical basis of any life that might be discovered. Moreover, such stains, when employed with a microscope, might permit the detection of small organisms that otherwise could escape notice.

PLANETARY EVALUATION

Of the nearby planets Mars offers the best chance of finding life. Although water and oxygen are scarce, the range of temperatures on Mars is moderate; and in fact, there are a number of terrestrial microorganisms which would certainly be able to survive in the Martian environment. Whether they could actively proliferate will probably depend on the local availability of moisture; obviously a map of Martian humidity would be a major objective of early exploration, as water may be the most prized mineral on the planet.

There is a long record of speculation on whether or not life exists on Mars. Much of the evidence has been at best only suggestive. Recent observations of the infrared reflection spectrum of Mars appear to indicate the presence of C-H molecules in the dark areas, materials whose abundance seems to vary with the seasons. An urgent objective is the further study of this phenomenon, and a probe to the near vicinity of Mars could provide important data unobtainable from earth.

Venus seems less likely to contain a recognizable form of life. Its surface temperature is apparently higher than the boiling point of water, according to indirect determinations from radio wave measurements. As in the case of Mars, however, it would be intemperate to make dogmatic conclusions until we have mapped the planet in more detail. In the case of Venus, with its dense atmosphere, a three-dimensional map of temperature and moisture is needed to indicate the zones where life might exist.

The moon, barren of atmosphere, appears unlikely to harbor life. But we know nothing of the conditions beneath its exposed surface.

The moon may be especially interesting as a gravitational trap for meteoroidal material accumulated from space over many eons, unaltered by the action of atmosphere. The earth is the most likely source of such material; and should terrestrial material be identified on the moon, it would furnish strong support for the hypothesis of panspermia. Mercury appears much like the moon physically. Its dark side, perpetually free from the corroding effects of the solar radiation, may be a far better repository of meteoroidal material than the moon, although its distance makes it a less likely target.

The large planets are too distant for examination in the near future, but they are not without interest. Conditions there would favor the accumulation of organic material. Compounds formed in the chemically reactive conditions of the upper atmosphere would descend to lower altitudes where they would be shielded by the atmosphere from decomposition by the solar radiation. It has been calculated that by now enormous quantities of carbon compounds have accumulated on the large planets.

Like all advances in human technology, space exploration may create new problems for mankind; and the biologist must be prepared to contribute toward minimizing them. One source of concern can already be anticipated. A consequence of our ability to send vehicles to other planets is the infection of these planets with terrestrial life. A probe, for example, could introduce microbial life onto a planet; and if there were nothing to limit growth, the organisms could occupy the entire planet in days or weeks. By the time a later probe was sent it would be too late to study the planet in its virgin condition, denying us an inestimable prize for the understanding of our own life and its origins. Moreover, we cannot exclude the possibility that such a catastrophe would have economic repercussions as well; it would be rash to predict too narrowly the ways in which undisturbed planetary surfaces might serve human needs. Also we must face the conceivable moral problems raised by the thought of our contaminating an already inhabited planet.

Fortunately, the responsible agencies in the United States and in the Soviet Union have recognized this problem and will insist that due precautions be taken on planetary missions. Vigorous methods of decontamination of spacecraft, principally by gaseous fumigation, are being developed. A conservative policy can do no harm, and caution need not preclude enthusiastic exploration.

By the next decade vehicles may be making round trips to the planets, and we must reckon with the bizarre possibility that the returning vehicles may infect earth with life from some other planet. Since the very existence of extraterrestrial life is still speculative, we can only surmise what such back contamination might mean. There could be new plant or animal diseases. If nothing more, we could imagine serious competition for some essential resource from a life form of extraterrestrial origin. The probabilities of these occurrences cannot be guessed. We must protect ourselves, however, by designing and sending automatic instruments that will return information without the peril of invasion of earth by some strange organism. In time, with more information on conditions on other planets, we will have a far better basis for deciding on the advisability of bringing materials from other planets to the earth.

For the biologist, space research is a gamble against high odds for higher stakes. The experiments are difficult to design, requiring years of preparation, and much of the data will be inconclusive. But there is the tantalizing prospect of finding answers to some of the oldest and most basic questions of biology. It can be argued that biology could gain more from space exploration than any other major field of science. At the same time, the biologist has a unique and grave responsibility to consider the implications of space exploration for human welfare.

SUGGESTED READING

ANFINSEN, C. B. *The Molecular Basis of Evolution.* New York: John Wiley & Sons, 1959.

COCCONI, G., and MORRISON, P. "Searching for Interstellar Communications," *Nature,* 184 (1959), 844–46.

HOROWITZ, N. "The Origin of Life," in *Frontiers of Science,* (ed). E. HUTCHINGS, JR. New York: Basic Books, 1958.

LEDERBERG, J. "Exobiology: Approaches to Life Beyond Earth," *Science,* 132 (1960), 393.

LEDERBERG, J., and COWIE, D. B. "Moondust," *Science,* 127 (1958), 1473.

MILLER, S. L., and UREY, H. C. "Organic Compound Synthesis on the Primitive Earth," *Science,* 130 (1959), 245–51.

OPARIN, A. I. *The Origin of Life,* trans S. MORGULIS, New York: Dover, 1953.

"Report of the Committee on the Exploration of Extraterrestial Space (CETEX)," *Nature,* 183 (1959), 925–28.

RUSH, J. H. *The Dawn of Life,* Garden City, N.Y.: Hanover House, 1959.

FLYING TELESCOPES

LYMAN SPITZER, JR.

Astronomy may be revolutionized more than any other field of science by observations from above the atmosphere. Study of the planets, the sun, the stars, and the rarified matter in space should all be profoundly influenced by measurements from balloons, rockets, probes, and satellites. Here I should like to dwell briefly on more distant parts of the universe and to outline the specific ways in which our knowledge of stars, interstellar clouds, and galaxies can be increased by use of the powerful new tools of space technology.*

Since exploration with spacecraft traveling out to stellar distances seems very far in the future, our knowledge of the stars and of interstellar material must be based primarily on the electromagnetic radiation which reaches us. Nature has thoughtfully provided us with a universe in which radiant energy of almost all wave lengths travels in straight lines over enormous distances with usually rather negligible absorption. In principle, useful information could also be obtained from energetic particles ejected, for example, by exploding stars. However, it is believed that interstellar magnetic fields, of order 10^{-6} gauss or more, deflect all charged particles except for the most energetic ones (energy much larger than 10^{15} electron volts) so that all trace of original direction is lost when a charged particle reaches the earth. I shall consider here only the electromagnetic waves as an astronomical tool.

The electromagnetic radiation, after traveling many billions of miles without hindrance, is strongly disturbed by the earth's atmosphere. Energy of most wave lengths (X-rays, ultraviolet, infrared) is entirely absorbed. Visual wave lengths penetrate but are slightly deflected by the changing thermal and turbulent

---
*This study was supported in part by Air Force Cambridge Research Laboratories.

structure of the air. The result is that stellar images are usually broadened to an angular width of about one second of arc. This is about the size of the diffraction image produced by a four-inch telescope; evidently a larger telescope cannot achieve its theoretical resolution unless it is placed above the atmosphere. A "flying" telescope is also required to measure radiation in the ultraviolet and other wave lengths which are absorbed by the atmosphere.

Since the contribution of space technology to astronomy can best be understood in terms of the questions that astronomers are trying to answer, the following section is devoted to the present research problems of galactic and extragalactic research. Subsequent sections discuss, first, the research that could be carried out with the higher resolution in visible light made possible by a flying telescope and, second, the questions that could be answered by measurement of radiation that does not penetrate the earth's atmosphere.

PROBLEMS OF ASTRONOMY

Outstanding research problems fall into four main areas which, together with solar and planetary studies, constitute modern astronomy.

1. *Constitution and evolution of stars.*—To understand the equilibrium of a star we must know its mass and chemical composition. Physical theory is then used to compute the equilibrium configuration, yielding values of the radius and luminosity. The measured values of these quantities then serve as a check on the theory and, particularly, on the chemical composition. Observed masses are obtained from the orbits of double stars, luminosities from the radiation received, and radii either from eclipsing binaries or from luminosities and spectroscopically measured temperatures. This field of research has had marked success in recent years in tracing how stars evolve and how the observed distribution of heavy elements can be explained by nuclear processes occurring in stars.

2. *Stellar atmospheres.*—The outer layer of a star is accessible to direct observation, and astronomers determine surface

temperatures and chemical compositions by interpretation of spectroscopic data. Detailed information on the abundance of elements gives information on specific nuclear processes occurring in various types of stars. More subtle analyses attempt to understand the complex structure of these atmospheres, including turbulence, magnetic fields, and radiation pressure. A detailed kinetic description is frequently possible, but the dynamic processes are still obscure. The uncertainty is greatest in exploding stars, novae and supernovae, where shock waves move outward at high velocities.

3. *Interstellar matter.*—The clouds of matter between the stars can be detected either by their absorption of starlight or by the radiation which they emit. In particular, measurement of atomic absorption and emission lines gives information on composition, densities, temperatures, and velocities of these clouds. Such information is still somewhat fragmentary because there are very few interstellar lines in the visible region of the spectrum. Also, small solid dust particles (grains) are detected by the continuous absorption of starlight which they produce. The limited data are consistent with the view that stars form from these clouds of gas and dust; as the stars grow old, much of the material is ejected back into interstellar space.

4. *Stellar systems.*—Measurement of the distances and velocities of the stars provides information on the form and internal motions of the galaxy and of the various subsystems within it. These data, plus radio astronomy observations on the 21-centimeter line of neutral hydrogen, have delineated the spiral arms of our own galaxy and have determined the relative motions of gas and stars. Data on different subsystems give important clues to the origin and evolution of the galaxy. Measures on other galaxies, which are apparently all moving away from us and from each other, give promise of indicating the nature of space, particularly its curvature, and possibly also the early history of the universe.

RESEARCH WITH HIGH RESOLUTION

The enhanced resolution possible with a flying telescope can

FIG. 1—Model of the orbiting astronomical observatory which the National Aeronautics and Space Administration will launch in 1964. It will lift telescopes above the atmosphere for observation of stellar radiations which are obscured at the earth's surface. The eight-sided satellite is 9.5 feet high, 6.5 feet in diameter, and weighs about 3,200 pounds (including instruments). The four paddles at the sides are covered with solar cells, which will provide the 350 watts of power required to operate the satellite and its experiments. (NASA photo.)

FIG. 2—Orbiting astronomical observatory in earth orbit. Two flight model observatories will cost approximately $23 million to develop and build.

be helpful in many astronomical problems. We consider here the research that could be done with two types of instruments, first a 36-inch telescope, with a resolution of 0.1 second of arc, and second, an instrument some ten times larger, with a resolution of 0.01 second of arc. An instrument of this first type is already under construction (Stratoscope II) for the Princeton University Observatory and is designed to be carried to high altitude by balloon. The orbiting astronomical observatory program of the National Aeronautics and Space Administration could also accommodate a satellite telescope of this sort.

The second instrument is much farther in the future. Because of flexure in a gravitational field such a large telescope must "float" in space. Its hypothetical high resolution probably could not be maintained if the instrument were placed on the moon, for example. One of the most interesting problems that such high resolution instruments could explore would be the surface details of the sun or of the planets. With the larger instrument, an object on Mars a mile in diameter could be resolved.

In the study of individual stars the chief importance of high resolution is in concentrating the stellar image in a smaller angle, making possible the detection of fainter stars. Thus a 36-inch telescope above the atmosphere could detect stars between 1 and 10 per cent as bright as the faintest stars detectable with the 200-inch Hale telescope on Mount Palomar, although much longer exposure times would be required. The reduced background of skylight for a satellite above the night airglow aids in the detection of faint objects. Hence stars of very low intrinsic luminosity whose very existence is still unknown could be investigated with these high-resolution telescopes. With a 360-inch telescope one could just begin to resolve the largest stars and might thus gain new information about the atmospheric structure of these objects. A two-mirror interferometer, with a separation of several thousand feet between the mirrors, could measure the radii of normal stars with great precision. High resolution also provides the possibility of separating very close double stars and thus extending measurements of stellar masses.

In studies of interstellar matter high resolution on gaseous nebulae would be of particular interest. The structure and dynamic evolution of these objects, which probably depend on the properties of an ionized gas in a magnetic field, are still a mystery.

It is in the field of stellar systems that high resolution would be most obviously useful. With investigations possible on intrinsically very faint stars one might hope to obtain an accurate inventory of the mass present in stars of all types; it is believed that much of the material in the galaxy is in the form of stars too faint to detect with present instruments. The apparent parallactic motion of the stars, caused by the earth's motion around the sun, could be used to obtain stellar distances for relatively remote objects—out to almost 1,000 parsecs (unit of measurement for interstellar space, equal to a distance having a heliocentric parallax of one second; 3.26 light years or 19.2 trillion miles) with a 36-inch and nearly 10,000 parsecs with a 360-inch. Apparent motions due to random stellar velocities across the line of sight could also be measured much more precisely. Crowded stellar subsystems within the galaxy could be resolved into separate stars for studies of their age and evolutionary history. Finally, our knowledge of the form and structure of external galaxies would be enormously increased. Fainter stars in these systems could be resolved and thus the distance scale of the external galaxies could at last be placed upon a firm footing. With fainter galaxies subject to observation and analysis, the extent of the universe might be definitely ascertained.

X-RAY, ULTRAVIOLET, AND INFRARED RESEARCH

While high-resolution observations are of comparable importance with observations over an extended wave-length range, the first astronomical satellites are being equipped for the latter rather than the former. This is because a significant increase in resolution can be achieved with balloon-borne telescopes at much lower cost than with satellites. Moreover, only relatively simple equipment is needed on a satellite for preliminary observations in wave lengths previously inaccessible, in contrast to the large, precisely guided and necessarily complex telescope required for high resolution.

For the same reason, ultraviolet rather than infrared detection equipment is likely to be launched with the first satellites. Much infrared research can be carried out from balloon altitudes. In fact, more work than is now under way could profitably be done with ground-based telescopes, looking in those few wave-length regions where the atmosphere is relatively transparent. The detection of X-rays requires altitudes higher than can be reached with balloons, and the equipment needed is relatively simple. Directivity is not easy to obtain, however.

Major astronomical programs could be carried out using wave lengths previously inaccessible. In the area of stellar internal constitution, determination of the luminosities of the hotter stars would be a great contribution to our knowledge. Those stars which are of particular importance because of their great brightness, recent origin, and relatively rapid evolution, radiate most of their energy in the ultraviolet. Hence the available measures of their visual luminosities provide only a poor indication of the total energy they radiate. Relatively simple measurements of the apparent stellar brightness of these hot stars at wave lengths between 1,000 and 3,000 angstroms would eliminate this major uncertainty in the theory of stellar structure and evolution. Similarly, measurements in the infrared would be important for cool stars, which radiate most of their energy at considerably longer wave lengths than those of visible light.

In the study of stellar atmospheres, measurements over an extended wave-length range would be helpful in analyzing the atmospheric composition. It is well known that stars are mostly composed of hydrogen and helium. The next most abundant elements are carbon, nitrogen, oxygen, and neon. The absorption lines produced by all these atoms in their ground states lie in the ultraviolet. Lines in the visible spectrum are absorbed by excited atoms, but the fraction of atoms which are excited in a stellar atmosphere is not certain, and hence the abundances determined from the line strengths are subject to substantial errors. Interpretation of the ultraviolet spectrum will be complicated by the enormous strength of the ultimate lines, and

Flying Telescopes

probable overlapping of lines, but in principle this additional information can increase our knowledge of stellar composition. Measurements of absorption lines in the infrared would also be useful, since the fraction of atoms excited to states very close to the ionization limit is better known than the fraction excited to intermediate levels.

A better knowledge of the complex physical processes occurring in stellar atmospheres could also be provided by measurements in unfamiliar wave lengths. This is particuarly true for exploding stars, such as novae and supernovae. Measurements at very short wave lengths, such as of X-rays and gamma rays, might be expected to yield fundamentally new information as to what is going on in these mysterious objects. Such measurements may even be of interest for normal stars, if extended coronae at million-degree temperatures generate large fluxes of these penetrating photons. The sun is known to possess such a corona and in more luminous stars this phenomenon may be much enhanced.

Measurements of eclipsing variable stars in the ultraviolet and infrared would also be a promising program. These stars provide reliable information on stellar masses and radii, as well as on atmospheric structure. If one of the two stars is substantially fainter than its companion, now only the brighter star can be detected, which drastically limits the information which can be obtained. If the fainter star is somewhat hotter, its spectrum will dominate in the ultraviolet and each star can be analyzed separately. Measurements of pulsating variables in the ultraviolet and infrared would be very useful in helping to separate light variations caused by changes in temperature from those caused by changes in radius.

INTERSTELLAR GAS RESEARCH

Our knowledge of interstellar clouds would be enormously increased by observations of ultraviolet absorption lines produced by these clouds in the spectra of distant hot stars. In interstellar space the atoms spend virtually all their time in the ground state, where atoms of all the abundant elements cannot absorb visible light; hence the only interstellar absorption lines of appreciable strength in the visible spectrum are produced by the relatively scarce elements, sodium and calcium. Measurements in the ultraviolet would make possible a direct determination of chemical composition and would yield much more information on the velocities and structure of these clouds, which are believed to be the birthplaces of new stars.

One of the most important constituents of the interstellar gas that might be discovered in this way is molecular hydrogen, which cannot be detected in small quantities with visible light. In view of the cosmic predominance of hydrogen, one might expect hydrogen atoms in space to combine, especially since the small solid particles should serve as an effective catalyst. Strong lines of hydrogen between 1,000 and 1,100 angstroms should be measurable with an ultraviolet spectrometer on a flying telescope. More detailed information is also desirable on the solid particles which observations show absorb starlight. Measurement of this absorption in the far ultraviolet and in the infrared should increase our knowledge of the composition and size distribution of these grains.

It has been suggested that some of the radiation observed from gaseous nebulae is synchrotron radiation, emitted by relativistic electrons in a magnetic field. Measures of this radiation in the ultraviolet and detection, possibly, of X-rays from these objects might yield fundamentally new information on the physical processes occurring in these magneto-fluids.

Another hypothesis that might be tested is the following: That surrounding each galaxy and in the vast spaces between galaxies there extends a very hot gas of highly ionized atoms. Possibly such a gas would be detected by finding the absorption line of quadruply ionized nitrogen or of quintuply ionized oxygen in the spectrum of a neighboring galaxy; these lines at 1,240 and 1,035 angstroms, respectively, are in a region of the spectrum that is convenient for detection and measurement. The measurement of emission from a receding galaxy in the strong ultimate line of hydrogen at about 1,216 angstroms should give information on the density of hydrogen surrounding a galaxy.

Evidently the general purpose in extending the wave-length

range of astronomical observations is to look for stars and atoms which cannot be detected with ordinary visible light. In a new adventure of discovery no one can foretell what will be found, and it is probably safe to predict that the most important new discovery that will be made with flying telescopes will be quite unexpected and unforeseen.

ROCKET PROBES

WILLIAM W. KELLOGG

A century or more ago the upper atmosphere must have seemed to most men remote and of little concern. Perhaps the northern lights attracted their admiration, shooting stars were familiar sights on a clear night, and the curious shiftings of the compass needle at certain times suggested that there were electric currents somewhere far above them. But there were few who attempted to explain these things.

Nevertheless, in the decades which separated the "First Polar Year" (1882-83) and the first sounding rocket fired in the United States at the White Sands proving ground (1945), a great many people turned their attention to the upper atmosphere. It is remarkable how well the conditions there had been described even before the first direct rocket or satellite measurements were made. In 1948 a Rand Corporation analysis of atmospheric conditions was published, giving tables of conditions out to 10,000 miles—though at that time only a few rockets had gone to 100 miles. The values were far from exact, but the general features were borne out quite well by subsequent measurements.

The deductions that made these estimates of conditions possible were based on a complex mass of data on meteors and their light-intensity distribution, airglow and auroral spectra, radio reflections, geomagnetic storm records, and acoustic propagation measurements. Every scrap of evidence that could be pieced together on the ground was used to forge a series of theoretical conjectures about a region where scientific instruments had yet to penetrate directly.

Then came the era of direct measurements with rockets (and later with satellites), and the roughly sketched picture of the upper atmosphere began to be filled in with details and a sense of perspective. There is an important lesson to learn from this brief look at the development of the picture that emerged. The upper atmosphere is no longer remote and unattainable, and anyone who can afford a sounding rocket can send instruments there. Soon scientists will be able to go there themselves, no doubt. But in our preoccupation with the rockets and their cleverly conceived instruments, let us not forget the many reliable ground-based sources of information, and the deductions that can be made from them. The time has come to compare notes, and to review the possibilities for combining rocket observations with indirect surface observations.

Another development that should be emphasized is the emergence of cheaper sounding rockets, bringing exploration of the upper atmosphere within reach of many countries. At the 1958 International Geophysical Year meeting on rockets and satellites in Moscow, four countries reported successful rockets: the United States, the Soviet Union, England, and Australia. Japan was working on one which proved itself shortly thereafter, and it was known that France had a successful sounding rocket which was not discussed at the IGY meeting. Canada, though not developing its own rocket, was host at Fort Churchill to the most extensive and successful series of firings of the IGY program. Now there are plans in various stages of completion for at least six more countries to fire sounding rockets, and a world-wide program is in the making.

Meteorologists are used to thinking of the atmosphere as a three-dimensional compressible fluid that is heated and cooled and moves in accordance with the conventional laws of fluid dynamics and thermodynamics. This is true in principle, even though the real atmosphere cannot yet be described accurately in mathematical terms.

As one moves higher in the atmosphere, one arrives at a region where fluid dynamics alone cannot account for the motions and the energy changes, since here free electrons and ions interact with the earth's magnetic field, and the atmosphere begins to act like a weakly ionized plasma. At this level (roughly the E region of the ionosphere) and above, the atmosphere is bombarded with solar ultraviolet and X-radiation and with charged particles from several sources. These radiations dissociate the molecules of oxygen (O_2), nitrogen (N_2), carbon dioxide (CO_2), and water (H_2O), and cause them to be ionized. The complex interactions that result are generally studied under the heading of "aeronomy."

Ever since the first balloons of Tesserenc de Bort, flown at the turn of the century, meteorology has pushed upward to seek a better understanding of the behavior of the atmosphere. Since World War II an international balloon radiosonde network under the guidance of the World Meteorological Organization has been making two simultaneous measurements of temperatures, pressures, and winds to about 20 kilometers on a routine basis. Some balloons go higher, and the United States Weather Bureau, on an experimental basis, has been able to construct synoptic (based on simultaneous observations) weather maps of the entire northern hemisphere for the 10 millibar pressure surface, at a height of about 30 kilometers. These were constructed with observations taken during the IGY, when a special effort was made to have high altitude balloon runs. However, 30 kilometers (or 10 millibars) seems to be about the upper limit to which balloons can be pushed on a routine basis. (The record for a very large experimental balloon is just over 40 kilometers.)

It has been possible, however, to get a good idea of the mean conditions at much greater heights. In 1951, one of the first models of the upper atmosphere was drawn, showing the mean temperature, pressure, and wind distribution up to 120 kilometers and from winter pole to summer pole. There was considerable guesswork in this model, since at that time there had been published only rocket observations at White Sands, New Mexico, and the rest of the picture had to be filled in by some acoustic measurements in Alaska and the Canal Zone sponsored by the Air Force, and by indirect ground-based observations of various kinds coupled with theoretical reasoning. Since then there have been many more rocket flights at virtually all latitudes and seasons, and a much more complete and detailed picture has now been drawn.

The scheme of the general circulation of the upper atmosphere is shown in Figure 1, which gives only the basic elements. The terms "warmer" and "colder" in the polar regions are with

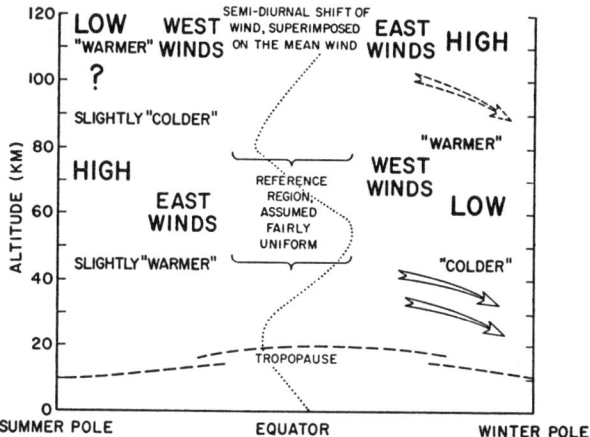

Fig. 1—Diagram of the mean wind, pressure, and temperature conditions in the stratosphere (roughly 10 to 50 kilometers altitude), mesosphere (roughly 50 to 80 kilometers), and lower ionosphere (above 80 kilometers) at solstice.

reference to temperatures at low latitudes where there is no seasonal change. The terms "low" and "high" refer in this simple model to pressure centers over the poles as they would appear on a weather map. Since the air in the northern hemisphere blows counterclockwise about a low pressure center and clockwise in the southern hemisphere, there is a west wind in both cases around a "polar low." By the same token, a high pressure center over the pole gives east winds in both hemispheres.

The air is generally colder in winter and warmer in summer below 60 kilometers, as one would expect from simple considerations of solar heating. At 80 to 100 kilometers the situation is just the opposite, and the polar air is colder than that at lower latitudes in summer and warmer in winter. In fact, the coldest spot in the entire atmosphere is over the summer pole at 80 kilometers, where it is about 170 to 180 degrees Kelvin on the average, as revealed by the series of IGY rocket firings from Fort Churchill during January and February, 1958.

This curious anomaly in temperature remains to be explained completely, but one theory advanced by this author credits the warming in winter to a release of the chemical heat of atomic oxygen as it is drawn downward over the winter pole and recombines. Atomic oxygen is known to exist above 90 or 100 kilometers, since at middle latitudes molecular oxygen is dissociated appreciably at this level by solar ultraviolet. If this suggestion proves to be correct, then the winter poles are the seat of a unique process in the atmosphere in which a large-scale release of chemical energy takes place.

These descriptions of the mean conditions below 100 kilometers are still inadequate to explain the day-to-day dynamic behavior of the upper atmosphere. There is clear evidence from serial rocket ascents at a number of places that the upper atmosphere is extremely variable at times, just as is the lower atmosphere—a fact which should not surprise us. In particular, in the winter and late spring there is a period of rapid breakdown of the cold low pressure vortex and this breakdown is accompanied by abrupt warming in the mesosphere and stratosphere. How this breakdown is triggered and how it takes place is one of the more intriguing problems of upper air meteorology. There is a hint from the Fort Churchill IGY rocket firings that the breakdown and sudden warming occurred first at about 65 kilometers (perhaps higher) and progressed downward to the 20-kilometer level in a period of a few days. Curiously enough, the south polar stratosphere appears to experience a more orderly and gradual breakdown, later in the spring than in the arctic stratosphere.

These polar breakdowns are but one feature of the upper atmosphere that interests meteorologists, and it is becoming clear that there are many complex interactions between the lower and upper atmosphere. A number of fairly strong correlations between disturbances in the ionosphere (usually measured by the degree of geomagnetic activity observed at the ground) and changes in tropospheric circulation have now been demonstrated, these correlations displaying a delay between the disturbance aloft and the change in the lower atmosphere. (This delay ranges from a few days to two weeks, depending on the lower atmosphere effect which is being studied.) At present, the possible linkages are unknown, and it is fair to say that we will not be able to trace the interactions between upper atmosphere (where solar changes are felt) and the lower atmosphere (where the weather takes place) until we have been able to study properly the atmosphere in between.

A study of this kind will require series of synoptic rocket ascents, so that a three-dimensional picture of the stratosphere, mesosphere, and lower ionosphere can be constructed and the day-to-day changes observed. The meteorological rocket network has been organized to do this in the United States. It is being run on a research basis and includes some nine potential launch sites, including Hawaii, Fort Greeley (Alaska), and Fort Churchill (Canada), as well as stations in the continental United States. According to its present schedule, it fires one rocket a day during every working day of January, April, July, and October (with some variations). The rockets go to about 60 kilometers effectively, observing winds, primarily, though some are equipped to measure temperature as well. The rockets generally used are the Arcas, which can carry some 5.5 kilograms to a height of about 70 kilometers (sufficient for a radiosonde and parachute), and the Loki II, a smaller rocket that has been limited for the most part to a three-kilogram "dart" containing radar chaff, giving winds alone from 65 kilometers down. Several manufacturers of solid propellant rockets have proposed improved sounding rockets in the $1,000 price range.

Plans are in a preliminary stage to start similar launchings in Europe. It is also hoped that Thule, Greenland, can be used for rockets. A look at the globe reveals that the United States Meteorological Network, Thule, and a few stations in Europe, would give a coverage of a major part of the northern hemisphere.

As one moves upward in the atmosphere the direct effects of solar ultraviolet and X-radiation and of charged particles pouring into the auroral zone become more pronounced. Each of these streams is absorbed as it passes into the atmosphere, and it causes the molecules to become dissociated into atoms and ionized. The molecule O_2 is appreciably dissociated starting at about 85 or 90 kilometers, and N_2 is partially dissociated above about 200 kilometers. These radiations also cause ionization, and the free electrons so created give the ionosphere its conductivity and account for its ability to reflect high- and low-frequency radio waves.

Because the ionosphere reflects radio waves, emits radiation (airglow and aurora), and has measurable electric currents, it was rather well understood even before rockets entered it. Nevertheless, the history of rocket experiments in this region is full of surprises. For example, the first hints of the very large fluxes of auroral particles (protons and electrons) into the upper atmosphere near the auroral zone came from State University of Iowa rockets fired to over 100 kilometers with Geiger counter detectors. These rockets were small solid propellant rockets carried to about 25 kilometers on balloons and then fired—the system known as a "rockoon." The discovery of these particles, even before the IGY, paved the way for an understanding of the Van Allen radiation belt, usually credited to the satellites.

Another surprise (to some) lay in the discovery that the "layers" of free electrons, which had been visualized as rather definite maxima in the electron density profile, were virtually non-existent when rocket measurements were made to an altitude of over 200 kilometers. Instead of peaks in the electron density profile it turned out that there was an almost steady increase of electron density with height during the day, on up to the level of the F_2 region maximum at about 300 kilometers. Above this level rockets fired to over 500 kilometers in the United States and the Soviet Union have shown a very gradual decrease in electron density with height, more or less as had been anticipated.

Through the years a series of brilliant observations of the composition of the ionosphere, made by American and Soviet scientists (and more recently by the British), have been made, in attempts to learn how dissociation and diffusive mixing take place. In these experiments, mass spectrometers of various designs are carried aloft to sample the air and to obtain relative abundances of neutral and ionized components of the atmosphere.

Another important aspect of rocket aeronomical studies has been the measurement of the flux of solar ultraviolet and X-radiation as a function of height. Since the atmospheric constituents O_2 and N_2 have characteristic absorption lines in the ultraviolet, and X-rays are absorbed more or less according to total mass of air (regardless of its state of dissociation), a careful measurement of the intensities of these solar radiations from a vertically moving rocket reveals the distribution of O_2, N_2, and total optical depth of atmosphere. This type of observation also tells a great deal about the sun, and has been extended to give ultraviolet fluxes from the cosmos as well.

Past rocket astronomy experiments, in addition to obtaining detailed solar spectra between 1,000 and 3,000 angstroms, have been able to demonstrate that the part of the solar spectrum that changes the most during a solar flare is in the soft X-ray region, at 5 to 50 angstroms. Recently, an important development was the measurement of the very intense line of singly ionized helium (He II) at 304 angstroms in the solar spectrum. It is so strong that it can account for a small but significant part of the daytime ionization of the lower F_2 region, since it is strongly absorbed by O_2 and N_2 above about 180 kilometers.

The recital of past achievements would make a much longer story, and these achievements are forerunners of many more experiments to come. A glimpse at the future possibilities of aeronomy research with rockets is contained in *Science in Space,* edited by L. V. Berkner and Hugh Odishaw. Perhaps the most important direction for future rocket research in the ionosphere is the use of combined rocket and ground-based measurements, since only by such experiments done many times will we be able to make proper quantitative use of the ground-based data. There is a large backlog of ground-based data, such as from ionospheric recorders, magnetic records, auroral photographs, radar reflections, and radio meteor echoes, but this must be related better to data from present efforts, since the interpretation is usually far from clear.

INTERNATIONAL ROCKETRY

In his first State of the Union address, President Kennedy invited "all nations—including the Soviet Union—to join with us in developing a weather prediction program . . . and in preparation for probing the distant planets of Mars and Venus." In jumping from weather prediction to planetary probes, the President passed over the middle ground where rocket probes are already being used internationally as one of the great hopes for better weather prediction.

A few key points in the complex picture of upper air rocket research include the following:

Rocket aeronomy has shown how the ionosphere reacts to solar radiation, and more particularly how changes in the sun (such as solar flares) cause drastic changes in our upper atmosphere. Among these changes are increased ionization (due initially to increased X-ray flux, and to increased charged particle flux at a somewhat later time) and heating of the upper atmosphere.

Rocket meteorology has shown that the lower ionosphere, mesosphere, and stratosphere are regions of change, and that changes at one level result in changes at other levels, though these interactions are imperfectly observed and scarcely understood. Statistical correlations between ionospheric activity (as measured by magnetic activity) and tropospheric changes have given meteorologists an unmistakable clue that there are indeed some links between the ionosphere and troposphere. To study these links we will need to have three-dimensional synoptic pictures of the intervening atmosphere, covering as much of a hemisphere as is possible.

Because of the need for such wide coverage the International Committee on Space Research (COSPAR) is sponsoring the development of a synoptic rocket network to observe the atmosphere properly. COSPAR has promulgated the International Rocket Weeks, in January, April, July, and October of each year, during which rocket firings are supposed to be co-ordinated around the world.

More to the point is the proposed extension of the Canadian–U.S. Meteorological Rocket Network, which is now struggling for its existence, to include truly co-operative rocket flights in Europe, eventually to be joined to the Canadian–U.S. network by a station at Thule. Here lies the possibility for a most fruitful international collaboration, and one which, like the radiosonde network, can do much to further our understanding of the complex dynamics of the atmosphere. It would be an important step toward the realization of the President's goal of an international program to improve weather prediction, since, clearly, prediction depends on knowledge.

THE EARTH AND NEAR SPACE

JAMES A. VAN ALLEN

Perhaps as recently as thirty years ago the working point of view of most physicists was that the planets moved about the sun in a near perfect vacuum, that the magnetic field of the earth fell off in intensity as the inverse cube of the radial distance from its center, and that the near vacuum of interplanetary space might be thought of as beginning at an altitude of some 1,000 kilometers above the earth. For certain purposes these views continue to represent a satisfactory approximation. Yet they ignore the foundations for a rich variety of physical phenomena. Indeed, the assertion that the density in interplanetary space is negligibly small is like the assertion that an atom is too small to be of any interest.

The development of the present state of knowledge of the near space environment of the earth has been a diverse and often murky process. The aim of this chapter is to sketch the current state of what is now a rapidly developing area of observation and understanding of the corpuscular radiations in near space—an area which is a blend of astrophysics, geophysics, and just plain physics.

The gravitational field of the earth presumably extends in a simple and reliable manner to an infinitely great distance, with its intensity declining as the inverse square of the radial distance. But it is the geomagnetic field, not the gravitational field, which dominates the environment of the earth in considerations pertaining to electrically charged particles. The geomagnetic field has a finite radial extension which is now known observationally to be about 15 earth radii. Terrestrial near space may then be loosely defined as a more or less spherical region around the earth of this size, the region being carried along with the earth in its orbital motion.

The modern era of knowledge of corpuscular particle physics

in the vicinity of the earth may very well be dated from the work of Chapman and Ferraro in the early 1930's. Indeed this work laid the conceptual groundwork in an important way for the now burgeoning modern field of magneto-hydrodynamics. These authors undertook to develop a physical model for geomagnetic storms based on the arrival in the vicinity of the earth of a cloud of tenuous hydrogen gas ejected from the sun. It was supposed that this gas, though neutral in large scale, was fully ionized, had a number density of the order of tens of ions and electrons per cubic centimeter, and had an electrical conductivity lying between that for metallic conductors and that for semiconductors. The interaction of such a body of conducting plasma with the geomagnetic field was shown to produce effects resembling those observed in actual magnetic storms. Despite the admitted incompleteness of the model it has served as a point of departure for many elaborations, including especially those of Hannes Alfvén. Electrical current systems in the plasma in which the earth is temporarily immersed and the precipitation of charged particles into the earth's atmosphere to make visible auroras emerge more or less naturally are features of those models.

Observational knowledge of the corpuscular environment of the earth has been advanced greatly during the past decade owing to the use of rockets, balloons, satellites, and space probes for the transport of experimental equipment to great altitudes; to the development of more refined and specific radio techniques and their increased use in polar regions, and to the much more assiduous study of the sun by optical and radio methods.

In 1953 and the following years, rocket observations off the coast of Greenland showed the arrival of energetic electrons in large intensities (approximately 10^6 to 10^8 electrons per square centimeter per second having energies in the range 10 to 100 kilo electron volts) by direct observation at altitudes of about 100 kilometers. Observations of this nature were later made in the Antarctic, by much improved rocket techniques at Fort Churchill during and since the International Geophysical Year (IGY), and by balloon techniques in a variety of Arctic and sub-Arctic locations. A general correlation of the arrival of large intensities of protons (approximately 10^5 per square centimeter per second per steradian) and electrons (approximately 10^{10} per square centimeter per second per steradian) having energies in the range from a few kilo electron volts to several hundred kilo electron volts and the occurrence of auroras as observed in traditional ways was established.

SATELLITE FINDINGS

The first United States satellite, Explorer I (February, 1958), opened up a large new era of knowledge concerning the corpuscular radiations within the earth's sphere of influence. Equipment designed to measure cosmic rays at a counting rate of about 30 counts per second was literally overwhelmed by rates exceeding 100,000 counts per second. Such rates would not have been surprising in high latitudes but the orbit of Explorer I encompassed only the region $\pm$ 33 degrees in latitude and up to 2,500 kilometers in altitude. The rudimentary observations were interpretable only as showing the existence of enormous intensities of electrically charged particles temporarily trapped in the geomagnetic field of the earth and within a region near the equatorial plane in which there had been no previous hint of their existence. The dynamic feasibility of such trapping of charged particles in the geomagnetic field was clearly implicit in the classical work of Stoermer, the Norwegian auroral theorist, beginning about the turn of the century. Retrospectively, it might be said that for over half a century the question had been open whether such theoretically possible orbits were or were not actually populated.

The experimental answer was a resounding Yes!

The effect was not a small one; the gross radiation exposure being comparable to, but of a different specific nature from that in the direct X-ray beam of a 50-kilo electron volt X-ray machine as used in ordinary laboratory work.

The early results have been confirmed and extended in many ways and a substantial literature on the subject now exists. The payloads of satellites Sputnik III, Explorer IV, Explorer VI, Explorer VII and the space probes Pioneer I, Pioneer III, Lunik I, Pioneer IV, and Lunik III have been devoted in part

The Earth and Near Space 121

to a study of the geomagnetically trapped corpuscular radiation (as it is now called). In addition, important work has been done by diverse equipment carried in smaller, relatively low altitude rockets.

Moreover, energetic electrons have been injected into the geomagnetic field by artificial means—bursting small atomic bombs at high altitude and the subsequent release of beta rays from the fission product. These experiments, called the Argus tests, were proposed by Nicholas Christofilos before discovery of the natural radiation but were not conducted until afterward (August–September, 1958). These injections of a known number of known energy particles at known positions in space provided a splendid test of many aspects of the dynamics of trapping of charged particles in the geomagnetic field. The immediately preceding hydrogen bomb tests in the Pacific (Hardtack series) and the Argus tests resulted in the first manmade auroras in the atmosphere and in many geophysical effects of global scale.

The discovery of the trapped radiation led to a number of enthralling questions: What is the composition of the trapped radiation? What is the absolute intensity and energy spectrum of each qualitatively distinct component? How do these properties vary with position in space? What are the lifetimes of particles in the trapping region? What geophysical effects do the particles produce—for example, auroras, airglow, geomagnetic effects, radio noise, atmospheric heating, and geonuclear effects? And last but not least, what is the origin of the observed particles? Or are there different origins for the various components and for the different regions of near space?

None of these questions has been satisfactorily answered. But in view of the difficulty of experimental investigations in this field, it is gratifying to realize that each of the questions can now be given a tentative answer, which is likely to resemble the truth of the matter.

The intensity structure of the trapped radiation exhibits two distinct and widely separated regions—the "inner zone" and the "outer zone." Each zone is approximately a figure of revolution about the (eccentric) magnetic axis of the earth. The maximum intensity of particle flux in the inner zone occurs on the magnetic equator and at a radial distance from the center of the earth of 10,000 kilometers (3,600 kilometers above the earth's surface). In meridian cross-section the inner zone has the shape of the broad face of a lima bean with the concave edge toward the earth. The complete inner zone is then like a loosely fitting ring or belt around the earth's magnetic equator. The magnetic forces so dominate the inertial ones that the two belts of fast moving, spiraling particles are effectively attached to the solid earth. Thus since the magnetic axis of the earth is tilted about 11 degrees to its axis of rotation, the radiation belts wobble up and down with a 24-hour period as viewed from a suitable astronomical vantage point. The lower edge of the inner belt is at an altitude of 400 kilometers over the east coast of South America and of 1,300 kilometers over the East Indies. The difference, 900 kilometers, divided by two, gives a simple and direct determination of the eccentricity of the magnetic center of the earth relative to its geographic center—namely, 450 kilometers.

The inner and outer zones are separated geometrically by a region in which the radiation intensity has a relative minimum. This region we have called the "slot." The slot lies along magnetic lines of force which reach the surface of the earth between magnetic latitude 45 degrees and 49 degrees depending on fluctuations in the outer zone; it crosses the equator at a radial distance of about 14,500 kilometers. The outer zone is a huge region with maximum radiation intensity at a radial distance (quite variable) of about 24,000 kilometers in the equatorial plane and with its outer limit at some 90,000 kilometers.

OUTER AND INNER ZONES

There are fluctuations by factors of 100 to as much as 1,000 in the intensity of charged particles in the outer zone. These fluctuations are associated with solar flares and with geomagnetic storms but detailed features of the correlation are not simple. One pattern of fluctuation is emerging as prominent, though not universal. Within about a day after a large solar flare and more or less coincident with the onset of a geomagnetic storm there is a marked depletion of the content of electrons of energy exceeding about 30 kilo electron volts in the outer zone. This depletion is accompanied by prominent auroras at latitudes some 10 to 15 degrees lower than those of the usual auroral zone. It is nearly certain that these low latitude auroras are a directly observable consequence of the dumping of previously trapped particles due to magnetic perturbation by the arrival of conductive solar plasma. Following the dumping process, the observed intensity of the outer zone increases characteristically, in approximately one day, to a value as great as, or often greater than, its pre-storm value. The intensity then "relaxes back" to a quasi-steady value. We believe that the post-storm recovery period is one of "local acceleration" of low-energy particles in a body of magnetically trapped plasma, from electron energies initially on the order of a few kilo electron volts or less to energies sufficiently great to be observable with equipment used thus far (threshold 30 kilo electron volts for electrons). There are a number of suggestions for local acceleration processes in the outer reaches of the magnetic field of the earth, none of which has strong experimental support as yet. It is hoped to undertake much improved observational study of the matter soon, employing detectors having a considerably lower energy threshold than any used in the past.

As intercepted at relatively low altitudes, say 1,000 kilometers, the outer zone exhibits its greatest intensity along a generally east-west band which at a longitude of 90 degrees west is centered on the Canadian–United States border.

There seems very little doubt that the outer zone owes its existence to the capture of ionized solar gas which sweeps by the earth in great clouds from time to time and is probably moving outward from the sun with greater or less intensity as a usual matter. This is the solar "wind." The zone of usual auroras lies along the magnetic shell which is the turbulent, unstable interface between the earth's trapping region and the interplanetary medium. The author believes that almost all auroras are intimately associated with the outer zone of trapped radiation and that most of the primary auroral particles which excite optical emission in the atmosphere have been trapped for greater or lesser periods of time in the geomagnetic field.

Trapped electrons drift around the earth toward the east; trapped protons drift to the west. The combined drifts constitute a westward flowing electrical current. There remain many obscure features in a full quantitative understanding of these processes, but it seems nearly certain that the outer radiation zone is the seat of a system of ring currents which make important contributions to the fluctuations of the magnetic field which have been noted from ground observatories. Satellite and space probe magnetometer measurements have directly detected variations in the magnetic field which can be plausibly attributed to this cause. Thus, the outer zone probably has an essential role in geomagnetic storms. The corresponding ring currents in the inner zone are apparently much weaker and, being relatively stable in time, constitute a nearly steady current system.

Time fluctuations in intensity are observed to be of progressively lesser relative magnitude as one moves inward from the outermost fringe of the outer zone toward the earth. During a three-year period the intensity in the inner portion of the inner zone near the equator has fluctuated by probably less than 20 per cent, though the definitive study of this matter is not yet completed. At the high latitude edges of the inner zone there are fluctuations on the order of a factor of two and in the slot, on the order of a factor of ten. These fluctuations also appear to be correlated with geomagnetic storms.

The inner zone appears to have an interior origin (from cosmic-ray-induced nuclear reactions in the atmosphere). The theory of this process leads to the expectation that the inner zone should extend outward to a much greater radial distance and a correspondingly higher latitude than has been observed. The reason for this great discrepancy is suggested by the nature of the observed fluctuation pattern. The postulated source of the inner zone is such a weak one that the mean lifetime of trapped particles therein must be on the order of ten years. A lifetime of this magnitude appears plausible if the loss of particles is simply due to quiescent energy loss and scattering in the residual atmosphere. Also the inner portion of the inner zone lies in a region of relatively strong geomagnetic field where the fluctua-

The Earth and Near Space 125

tions are slight. But as one goes outward, the fluctuations of the field rapidly become more important and the influence of such fluctuations is evident in the fluctuations of the particle intensity. Hence, the rapid decline of the intensity in the inner zone beyond a radial distance of 10,000 kilometers is probably due to an enhancement of the loss mechanisms.

The outer zone illustrates the matter in a much more striking way. The inner boundary of the outer zone corresponds to the complementary aspect of the same line of thought. Plasma from the sun intrudes into the magnetic field of the earth to different distances depending upon the kinetic energy of charged particles per unit volume. The inner boundary of the outer zone thus corresponds, presumably, to a sort of maximum depth of intrusion averaged over a suitable time interval. The slot between the two regions is then seen to be a consequence of the two complementary tendencies which arise from the same basic cause. The qualitative aspects of this argument apply to any other magnetized planet, but the quantitative features are dependent on the magnetic moment of the planet, on its radius, and on the structure and extent of its atmosphere.

The inner boundary of the inner zone is quite satisfactorily understood as being due to the greatly enhanced rate of energy loss and scattering of particles whose oscillatory trajectories dip into the increasingly dense atmosphere which occurs at these altitudes. The lower boundary of the outer zone at high latitudes may be understood similarly but the boundary is less well defined owing to a more or less continuous state of fluctuation.

COMPOSITION AND ORIGIN

One of the early findings with Explorer IV in July–August, 1958, was that the radiation in the equatorial region was much more penetrating than that at high latitudes. The existence of two distinct zones was suggested by the data at that time, though conclusive evidence was not obtained until the flight of Pioneer III in early December, 1958.

Meanwhile the suggestion of solar origin for the trapped radiation had been supplemented by the suggestion that the following process might contribute significantly. Cosmic ray particles produce nuclear disintegrations in the earth's atmosphere when they collide with nuclei of oxygen and nitrogen. Among other reaction products are neutrons having energies in the range from a few million electron volts to hundreds of million electron volts. After moderation in the atmosphere a fraction of these move outward from the atmosphere along trajectories not influenced by the geomagnetic field. A small fraction of 1 per cent

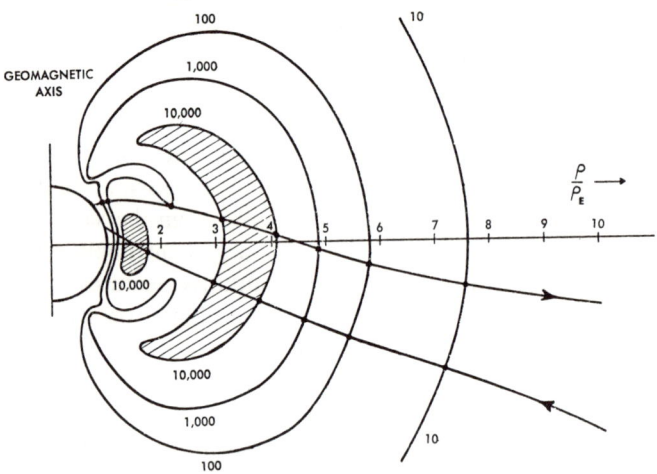

Fig. 1—Radiation intensities plotted with geomagnetic axis of the earth as axis of symmetry. Heavy, arrowed, non-symmetric line represents the trace of the Pioneer III probe, which recorded the data going outward and returning to earth. Lune-shaped curves represent inferred contours of radiation intensities around the earth. These findings were confirmed for space near the earth by Explorer IV. More recent satellite and probe data is altering the picture somewhat, and it is now thought that the belts fluctuate. Distances on the axis are given in earth radii; figures associated with contours represent radiation intensity in counts per second.

of them undergo radioactive decay within a flight path of several earth radii (mean lifetime against beta decay at rest, 12 minutes). The charged decay products are protons, having an energy spectrum similar to that of the parent neutrons, and electrons having an energy spectrum similar to the slow neutron decay spectrum (upper limit 780-kilo electron volts) but some-

The Earth and Near Space

what shifted toward higher energies. Of the diverse particles which compose the cosmic ray "albedo" of the atmosphere, neutrons are the best qualified to contribute to the trapped radiation—their delayed decay results in the injection of charged particles into the magnetic field at sufficiently high altitudes that the ensuing spiral oscillatory motion of these particles can often be confined between magnetic mirror points well above the dense atmosphere. The possibility that the decay products of secondary cosmic ray neutrons might populate trapped orbits in the geomagnetic field had been realized some months earlier by Nicholas Christofilos and described in various classified reports in connection with his proposed experiments for the artificial injection of electrons into the field.

The most persuasive early evidence for the quantitative validity of the neutron albedo hypothesis came from the successful recovery of a packet of nuclear emulsion flown into the lower edge of the inner zone which showed that under a relatively heavy shield (6 grams per square centimeter of material) there were indeed protons of measurable intensity up to energies of at least 600 million electron volts. Moreover, they measured the energy spectrum of the protons from 75 to 600 million electron volts and found that this spectrum had a plausible resemblance to that expected from albedo neutrons. This work has been repeated and refined more recently and better estimates of the source function, of the expected spectrum, and of the lifetimes of trapped protons have been made. Relying on the emulsion identification and on the quantitative intensities from Explorer IV and Pioneers III and IV, the author estimates that in the heart of the inner zone, the absolute omnidirectional intensity of the protons of energy greater than 40 million electron volts is 2×10^4 per square centimeter per second, with an uncertainty of a factor of two.

But in spite of the fact that energetic protons are the trademark of the inner zone, the intensity of electrons is very much greater there. Thus, we estimate the omnidirectional intensity of electrons of energy greater than 600 kilo electron volts as 1×10^7 per square centimeter per second in the heart of the inner zone (uncertain by a factor of three) and the intensity of electrons of energy greater than 20 kilo electron volts about two orders of magnitude greater. The electrons in the inner zone may also be supposed to be due to neutron albedo, but the theory of the loss of energy of trapped electrons is in a far less satisfactory state than that for protons, and little confidence can be placed in this supposition, though it does remain the most plausible one on qualitative grounds.

The outer zone is distinguished by the absence of energetic protons. The intensity of protons of energy greater than 75 million electron volts in the heart of the outer zone is less than one particle per square centimeter per second. Their absence is attributable to the simple incapability of the geomagnetic field to retain them in trapped orbits, especially when the field is perturbed from time to time by the arrival of solar plasma. Almost no observational information is yet available on the intensities of protons of energy less than 60 million electron volts in the outer zone.

In the heart of the outer zone the intensity of electrons of energy greater than 30 kilo electron volts is typically of the order of 10^{10} per square centimeter per second and of electrons of energy greater than 200 kilo electron volts of the order of 10^7. Massive fluctuations occur as outlined earlier.

No convincing proposal of anything other than solar plasma as a source for the outer zone has been made. But the detailed mechanism for producing the observed energy spectrum remains obscure. The nature of this mechanism is perhaps the most interesting unsolved problem in the subject of the trapped radiation, though the reader will be well aware by this point that a considerable diversity of other unsolved problems remains.

There is a glaring ignorance of the intensities of low energy protons and electrons in the outer zone. These components of the outer zone are very likely of dominant importance in producing geophysical effects, that is, heating of the atmosphere, auroras, airglow, and magnetic storms.

THE SUN

LEO GOLDBERG

Before 1940, observation of the sun was a major activity at only a handful of observatories throughout the world. Solar physics has since been expanding at an almost explosive rate; although exact figures are hard to come by, it is probably no exaggeration to say that about 100 new optical instruments for the observation of the sun, and an equal number of radio telescopes, have been put into operation in many countries during the past fifteen years.

Reasons behind this fantastic growth are no different from those that would cause any branch of science to expand: inherent importance of the subject coupled with a sudden implosion of new instrumental techniques and theoretical ideas. The end result is that knowledge of the sun is accumulating at a much faster rate than at any time since it was first viewed through a telescope by Galileo in 1611. Indeed, solar physics has not experienced such a rapid accession of new information since the founding of the Mount Wilson Solar Observatory by George Ellery Hale in 1904. Hale's invention of the solar tower telescope, of the spectrohelioscope and of the spectroheliograph (also invented by Deslandres at about the same time) initiated what was then a new era in solar physics, laying the groundwork for modern investigations into the physical nature of the sun.

The investigation of the sun was no less important in 1904 than it is today, and indeed the reasons given by Hale for the establishment of the Mount Wilson Observatory are the same ones that would be considered important by astronomers in 1961. As Hale put it in his first report as director, the purpose of a solar observatory is "the investigation of the sun as a typical star, in connection with the study of stellar evolution; and as the central body of the solar system, with special reference to possible changes in the intensity of its heat radiation, such as might influence the conditions of life upon the earth."

It is hardly necessary to elaborate on Hale's statement, except to point out that the second nearest star is about 250,000 times farther away than the sun and that events on the sun often lead to dramatic repercussions on the earth, such as the blackout of long-distance radio communications, brilliant apparitions of auroras, and the generation of magnetic fields in the earth, and of electric currents in the atmosphere. It also seems that fluctuations in ultraviolet and corpuscular radiation from the sun must somehow affect the weather although exactly how is yet to be discovered. The very recent discovery that the hoped-for venture of man into interplanetary space will expose him to intermittent and at present unpredictable blasts of lethal cosmic radiation from the sun has added a new element of urgency to the practical aspects of solar research.

THE JOB FOR TODAY

The immediate task of solar physicists is to assemble a complete and detailed picture of the nature and structure of the sun by interpreting observations of its radiation in the light of physical theory. Without such a picture, we cannot hope to answer the great questions of the sun's age and evolution, nor can we comprehend the nature and cause of the violent disturbances that reach out across 93,000,000 miles of interplanetary space to influence human activity on the earth.

It would seem that present science and technology should be able to acquire a complete description of the physical nature of an object as large and bright as the sun in a fairly straightforward way. But aside from the fact that we do not know enough about the behavior of hot, magnetized gases to interpret fully the things that we do see, the observation of the sun is impeded in two major respects—one, inherent, and impossible to overcome; the other, being eliminated by the introduction of space vehicles.

The first difficulty is that the solar gases are exceedingly opaque and therefore the radiation that reaches us emanates

The Sun

from a thin and highly tenuous outer envelope that contains only about one part in ten billion of the total solar mass.

The second limitation is inflicted by the earth's atmosphere, which distorts fine details on the solar surface and entirely obscures both ultraviolet and X-radiation at one end of the electromagnetic spectrum and long radio waves on the other.

Penetration of the solar interior is beyond the realm of present possibility, even though rocket probes may be sent to explore the outer periphery of the sun in the foreseeable future, but we can now look forward to observing the full range of the sun's electromagnetic spectrum and thereby to secure advantages on a scale comparable with those that resulted from the invention of the telescope. Before describing some of the steps that are being taken to exploit this opportunity, I shall indicate briefly the lines along which observations from space vehicles are expected to advance our understanding of the sun.

WHAT WE NOW KNOW

In relation to other stars, the sun (see Fig. 1) is average in size, mass and luminosity. For example, if it were located at a distance of 30 light years, it would be just plainly visible to the naked eye as a yellowish star of about the fifth magnitude. Its mass of 2×10^{33} grams and its radius of about 700,000 kilometers are both well established; however, the radius refers to the rather sharp boundary that the sun possesses when it is observed in visible light and not to its highly rarified outer envelope of gases which extends far out into interplanetary space, perhaps to the earth and beyond.

Despite the inaccessibility of the solar interior, a great deal can be inferred about its structure from its mass, luminosity, radius, and chemical composition, and from its obviously stable condition. Thus, its central temperature must be in the vicinity of 15,000,000° centigrade and its central density about seventy times that of water. At such high temperatures, thermonuclear processes readily convert hydrogen into helium at a rate sufficient to liberate the energy the sun is observed to radiate, and it is believed that the sun has been generating energy in this

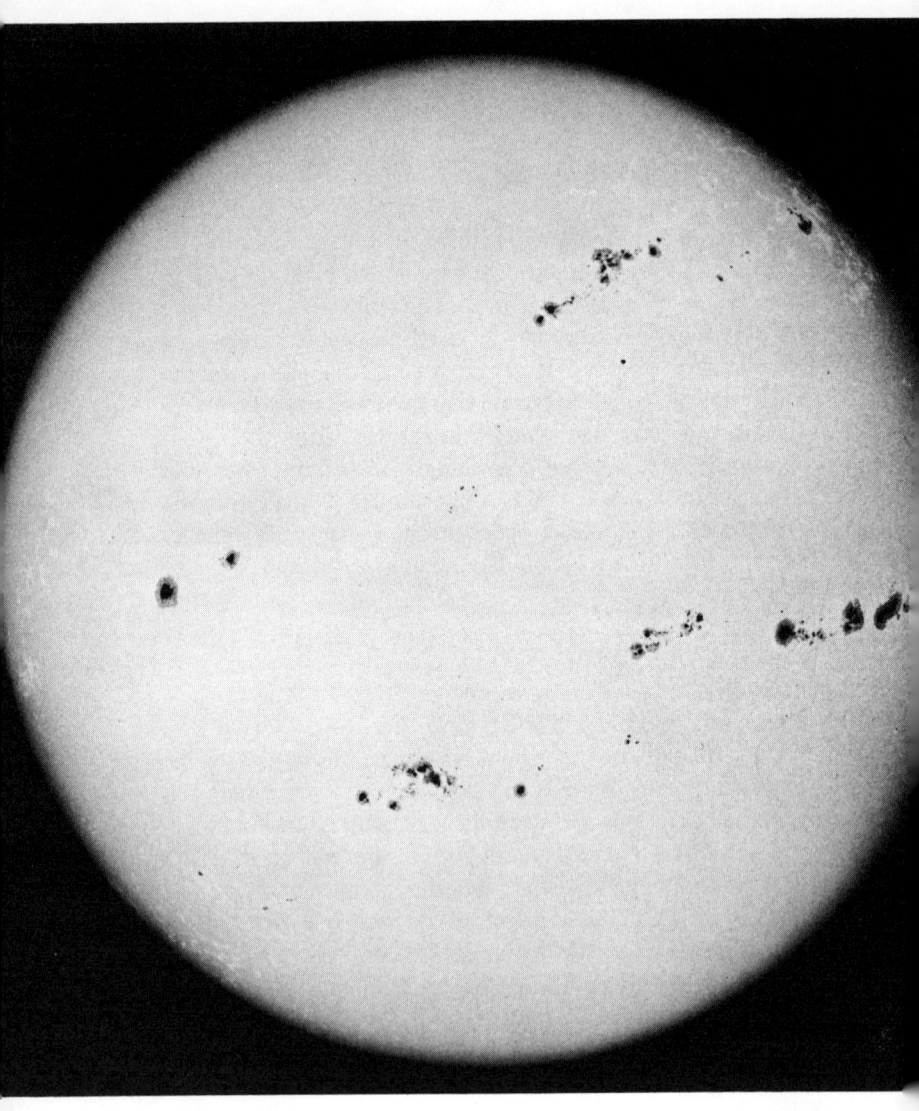

FIG. 1—Photograph of the sun's visible surface at sunspot maximum, December 21, 1957. (Mount Wilson and Palomar Observatories.)

way and at about the same rate for the past five or six billion years. Furthermore, since about 70 per cent of the sun's present mass is in the form of hydrogen, it has ample fuel to carry on for many billions of years in the future. Both the temperature and the density decrease steeply outward from the center to a layer a few hundred kilometers thick, called the "photosphere," which defines the visible surface. The gases at the bottom of this layer are completely opaque to visible light, whereas at the top they are virtually transparent. Since the angular thickness of the layer is but a fraction of a second of arc, the sun appears to have a sharply defined edge and its visible light seems to emerge from a surface, even though in reality it comes from all depths of a layer in which the temperature varies from about 4,500° to 8,000° centigrade and the density from one 100-millionth to one 2-millionth that of water.

SUNSPOTS AND GRANULATION

The photosphere is the best studied and best understood region of the solar atmosphere, primarily because most of its radiation is in the form of visible light and its observation is much less hampered by the earth's atmosphere than that of the sun's higher layers. The most spectacular phenomena observed in the photosphere are the *sunspots,* regions some 1,000° or 1,500° cooler than the normal surroundings, which occur over a large range in sizes up to nearly 100,000 kilometers in diameter (see Fig. 2). Sunspots are characterized by strong magnetic fields, up to several thousand gauss for the largest spots, and, as is well known, the number appearing on the sun at any one time varies according to an approximate cycle which also characterizes other types of solar activity. Although sunspots have been the object of intensive study since the time of Galileo, the observations have not yet been synthesized to provide even an accurate description of their physical nature, let alone the mechanism of their origin and development.

Apart from the sunspots, the photosphere is also the scene of a very widespread and continuous kind of activity known as granulation, which takes place on such a small scale that it is exceedingly difficult to observe from the ground through the earth's turbulent atmosphere, except with very careful observational techniques, but which has been revealed in exquisite detail in the solar photographs obtained from an unmanned balloon at 80,000 feet by Martin Schwarzschild. These photographs (see

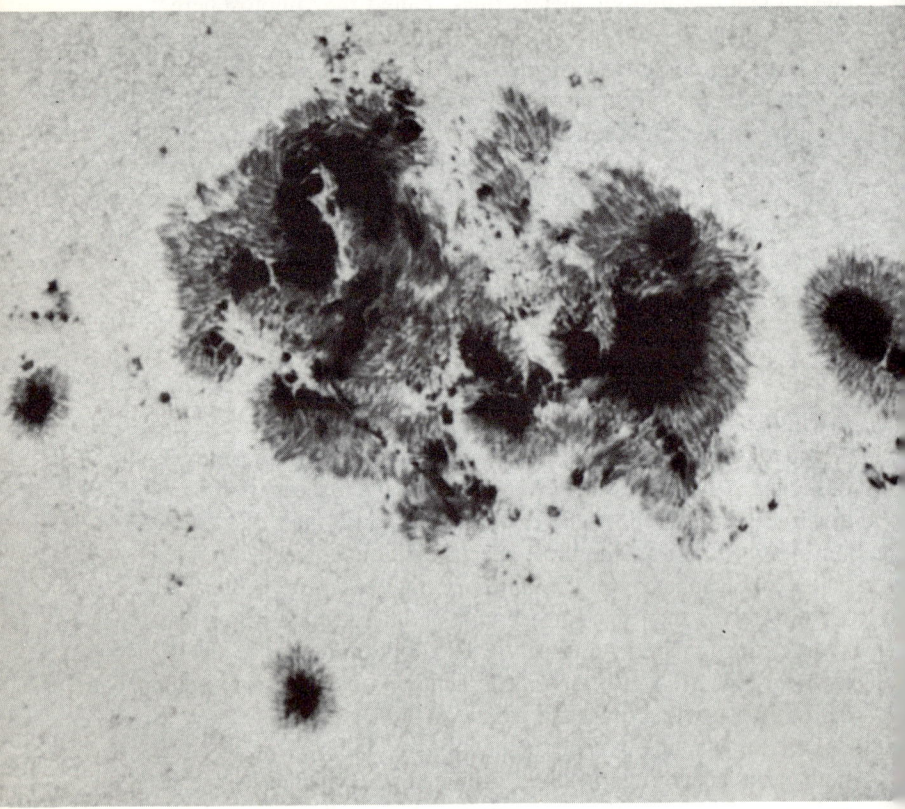

Fig. 2—Large sunspot group of May 17, 1951.
(Mount Wilson and Palomar Observatories.)

Fig. 3), as well as other observations stretching over the past century, show the sun's surface to possess a granular structure consisting of small bright regions about one second of arc or 800 kilometers in diameter, separated by narrower dark lanes which are about 50° to 100° cooler than the bright granules. This

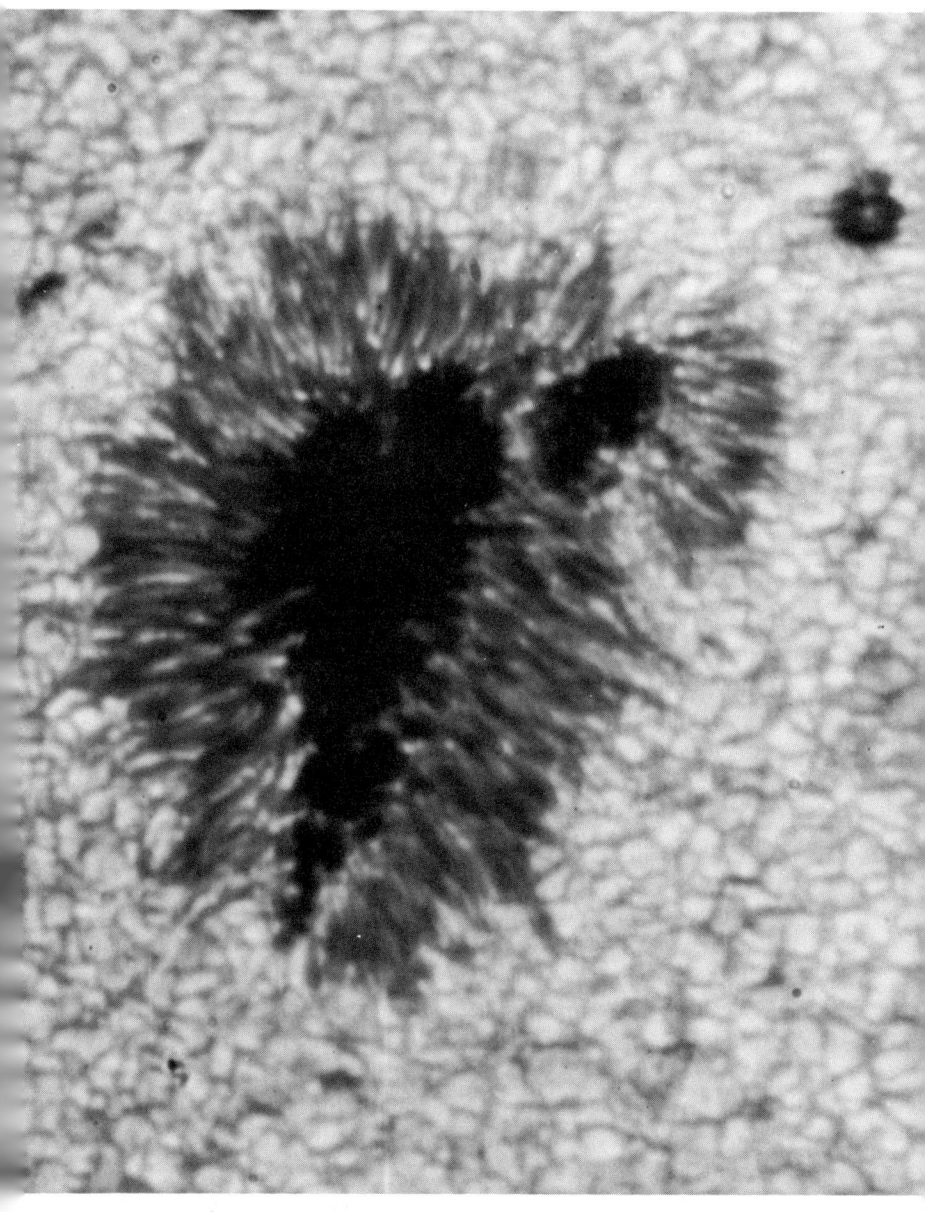

Fig. 3—Solar granules photographed August 17, 1959 from high-altitude balloon. (Project Stratoscope of Princeton University.)

rather elusive aspect of the sun's appearance turns out to be the key to much of the drama that occurs in the high layers of the sun's atmosphere above the photosphere, revealing the existence of a mammoth system of convection and circulation.

Throughout most of its interior, the sun exists in a very quiet state. The energy generated in the deep interior makes its way slowly outward to the surface by radiation rather than by the mass motion of the hot gases. However, in the outer 10 or 20 per cent of the sun's radius, the conditions are such that the rapid outward flow of energy cannot be handled by radiation alone and therefore the energy is carried upward by moving columns of gas, in short, by convection. The bright granules are undoubtedly the tops of violently churning bubbles of hot gas whereas the darker lanes must signify downward-moving matter after it has cooled. The motions are greatly complicated by the fact that the solar gases are magnetized in a highly irregular pattern; the usual fields do not exceed a few gauss but may attain values as high as 100 gauss in local regions, even in the absence of sunspots. At the top of the photosphere, the upward-moving gases have speeds in the neighborhood of 1 kilometer per second, but higher up the motions are accelerated and appear to culminate in a system of supersonic jets with speeds up to 30 kilometers per second.

The granulation causes an enormous flood of mechanical and magnetic energy to be poured into the sun's outer layers where it is dissipated into heat much faster than the gases can radiate it away. Consequently, the temperature rises so rapidly that it reaches a value of about $1,000,000°$ over a distance of about 20,000 kilometers. This transition layer of the solar atmosphere is called the "chromosphere," from its rosy appearance when observed at eclipses. The very high temperature region above it, known as the "corona," can be seen during solar eclipses to extend for many millions of kilometers into interplanetary space and, as Sydney Chapman has suggested, it may even reach the orbit of the earth and beyond.

Most of the exciting and spectacular stormlike disturbances that rage in the solar atmosphere occur in the chromosphere and corona. The totally unexpected discovery, by Edlén in 1940, that the corona is a high-temperature plasma—which repeatedly has been backed up, notably by radio astronomy observations—was the single most important stimulus to postwar solar research. It must be admitted, however, that the detailed physical nature of the chromosphere and corona and of the events that take place within them is still largely shrouded in mystery.

For example, neither the exact mechanism that heats the chromosphere and corona, nor even the rate at which the temperature rises, are known with any certainty, and the plain truth seems to be that different kinds of equally good observational data often yield conflicting and contradictory interpretations. There may be one or more still undetected physical processes occurring in the chromosphere and corona, which, when discovered, will resolve many unsolved puzzles.

While frequency of violent activity in the chromosphere and corona is correlated with prevalence of sunspots according to the 11-year cycle, individual storms may erupt at any time. The most spectacular and catastrophic of these disturbances is the solar flare—a great flood of energy, sometimes amounting to more than one billion kilowatt hours, released in a relatively small region of the chromosphere during a short space of time, a few minutes to a few hours. Flares have been observed on the sun for more than 100 years, and classically their appearance has been signaled by a sudden increase in the visual brightness of the affected region. More recently, it has been established that flare regions also emit intense bursts of radio waves which signify the ejection of high speed electrified particles traveling at speeds ranging from a few hundred kilometers per second up to several tens of thousands of kilometers per second. The earth is protected from most such onslaughts by the shield thrown up by its magnetic field, but particles with sufficiently high speeds often penetrate the shield and are received on the earth as solar cosmic rays. Although slower moving particles are deflected by the earth's magnetic field, they still incite disturbances in the high atmosphere of the earth—displays of auroras, and surges of electrical currents which alter the earth's magnetic field, sometimes reducing long-distance telephone and telegraph communications to chaos.

Recent brilliant experiments of Herbert Friedman and his associates at the Naval Research Laboratory, in which instrumental high altitude rockets were launched during several bright solar flares, have revealed that the flares also radiate X-rays with very high intensity, and increase the output of ultraviolet radiation. Impact of X-rays upon the earth's atmosphere increases the electrification of the ionosphere so much that radio waves transmitted from the surface of the earth are absorbed instead of being reflected as they normally are, occasionally causing abrupt blackouts of long-distance radio communication. It is therefore established that flares radiate strongly throughout the entire electromagnetic spectrum and with particularly high intensity at the extremes of short and long wave length which are blocked by the earth's atmosphere. Although occasional glimpses of these hidden radiations have been secured from high altitude rockets, their full and detailed study requires continuous observation of the sun on a 24-hour basis, which can only be accomplished with the aid of one or more satellite observatories.

An important beginning was made with the launching in the spring of 1960, of a small, "piggy back" satellite, simultaneously with the United States Navy's Transit II-A navigational satellite. This satellite was instrumented by Dr. Friedman for the observation of solar X-rays and ultraviolet radiation in selected wavelength bands, and during its lifetime of nearly a year it continuously sent accounts of solar activity back to the earth by radio.

The equipment aboard this first orbiting solar observatory was necessarily small and modest, but more advanced versions are being prepared for launching by the National Aeronautics and Space Administration. The NASA program is built around the so-called "S-16" and "S-17" satellites, designed and constructed by Ball Brothers Research Corporation of Boulder, Colorado.* The weight of the satellites, about 375 pounds, is large enough to carry devices to keep the solar instruments aimed toward the sun, thus allowing detailed study of small regions of solar activity. A number of groups in the United States are pre-

---

*The S-16 satellite was launched in March, 1962. Both the satellite and all of the on-board experiments have performed with outstanding success.

Fig. 4—Model of the S-16 orbiting solar observatory. The satellite will be stabilized and sun-oriented. Experiments aboard S-16 and S-17 will investigate a broad spectral range of solar radiation. The wheel-shaped section is 44 inches in diameter, and the satellite weighs about 441 pounds. The stabilized, fan-shaped section on top of the wheel carries two spectrographs, which will point continuously at the sun, and a solar cell array to provide electrical power for the satellite, which will have an estimated lifetime of six months. (Ball Brothers Research, Boulder Colorado photo.)

paring experiments for flight in the satellites, among them, NASA's Goddard Space Flight Center, the Naval Research Laboratory, and several universities—California, Colorado, Harvard, Minnesota, New Mexico, and Rochester.

The experiments to be carried by these satellites will still be relatively crude, but one can visualize, in the not too distant future, a very large orbiting solar observatory, weighing several thousand pounds, containing solar equipment comparing favorably with that of existing ground-based observatories and doing its job in response to radio commands from the earth.

Now the observation of the sun from an earth satellite represents an extension of techniques long employed at ground-based observatories, albeit to a vastly expanded range of observation. A further exciting possibility is that of sending instrumented probes into at least the outer extensions of the solar atmosphere for direct sampling of physical conditions, especially for measurement of the density of charged particles and magnetic fields. A start in this direction was made over a year ago with the voyage of the Pioneer V probe, which performed such measurements about twenty million miles closer to the sun than the orbit of the earth. The sun's great heat presents a problem at very close distances, but the use of highly refractory materials and advanced heat shields may permit probing instruments to get within a few million miles. The close co-ordination of measurements from probes and observations from satellites should revolutionize our knowledge of solar storms and of their impact upon the earth.

It must not be assumed that orbiting solar observatories and solar probes will cause ground-based observatories to become obsolete in the foreseeable future. Many studies, especially those dealing with visible light and short radio waves, can be carried out as well or better from the surface of the earth, while still others are ideally suited to balloons. In view of the enormous cost of space activities, astronomers have a very special responsibility to employ space platforms only for those experiments which cannot be done as well at their home bases. Inevitably, the results obtained from space vehicles will greatly stimulate new investigations from the ground and thereby accelerate the drive toward full understanding of the central body of the solar system which man is about to explore.

THE MOON AND PLANETS

GERARD DE VAUCOULEURS

The dawn of the space age has suddenly focused attention on a minor field of physical astronomy—the physics of the moon and planets, a field relatively neglected since the turn of the century.

Indefinite and often contradictory answers offered by the astronomer to questions raised by the technologist in planning interplanetary craft and direct exploration of the moon and planets, have exposed the great poverty of knowledge of physical conditions even on the nearer planets, the unavoidable result of a long period of neglect. The post-sputnik rash of theories about the moon and planets by all sorts of physical scientists with little or no prior knowledge of the subject has compounded the confusion and, incidentally, has done nothing to erase the stigma of ill repute earned for the subject by the overenthusiastic activities and extravagant claims of some famous amateur astronomers of a previous generation.

It is imperative to reconsider seriously the whole field of lunar and planetary research, for astronomers to accept it as a bona fide and significant subject worth serious consideration, for others to enter it without the hoopla and extravagance that characterized the sudden and belated discovery of outer space by the public. The exploration of the moon and planets is the ideal meeting area for a great variety of scientific disciplines; the astronomer has the techniques and experience required to secure the observational data and to provide the background information required by others to interpret these observations, but he lacks the specialized knowledge in meteorology, geophysics, geology, mineralogy—and perhaps biology—that is needed to advance detailed understanding of the phenomena disclosed by the astronomical observations.

There are many special reasons to be interested in the study and exploration of the moon and planets—the professional explorer seeks another Everest or Antarctica to conquer; the geologist is curious about unearthly rock formations; the astronomer wants to know more about planetary masses or diameters. Transcending these particular interests, two problems stand out as the most challenging and promising: one, the possibility that direct sampling and probing of the surface and crust of the moon may help discover clues to the origin, mode of formation, and prehistory of the solar system; the other, of even greater interest, is the probability that direct exploration of Mars may place us in contact with extraterrestrial forms of life whose study may help solve the problem of the origin and evolution of life under different planetary environments.

OBSERVATION TECHNIQUES

Even before direct exploration of the moon and planets by human visitors or by remote controlled sensors, many techniques can be used for physical studies of these celestial bodies from the surface of the earth and from stratospheric and space stations. The possibilities of visual, photographic, and photoelectric observations of planets and their atmospheres from existing observatories have not yet been exhausted. Such observations would be much more effective if performed from stratospheric balloons floating above the lower turbulent layers of the earth's atmosphere. These observations lead to geometric determinations of the diameter and oblateness of the globes, of their periods and axes of rotation, and of "geographic maps" from planetocentric coordinates of surface markings. Virtually all these elements must be improved if our future space explorers are not to find themselves stranded in uncharted territory. Photometric and spectrophotometric observations of the reflected solar light as a function of phase angle, the angle between the direction to the sun and the direction to the earth as seen from the planet, give fundamental information on the structures of the lunar and planetary surfaces and especially of planetary atmospheres, particularly such characteristics as chemical composition, gas density and atmospheric pressure, and particle sizes.

Photoelectric observations of the occultations of the stars by the planets are a powerful means of determining the structure (total pressure, density gradient, temperature, etc.) of the upper atmospheres. This penetrating technique has only recently been put into use and could be developed to secure information otherwise attainable only by direct penetration of planets' atmospheres.

Polarimetric observations have been made mainly visually in the past and should be extended both to the infrared and the ultraviolet. The polarization of light reflected by the planets is a subtle and penetrating indicator of the nature and structure of their surfaces and atmospheres. To derive full benefit from this technique, a substantial supporting program of laboratory measurements on terrestrial samples is required.

Spectroscopic observations, mainly photographic in the past, can now be advanced by photoelectric recording. This is the most direct technique, short of sampling *in situ,* for the determination of the chemical composition of planetary atmospheres, especially if such observations could be performed from satellites or balloons above the absorbing layers of our own atmosphere. Analysis of line intensities in absorption bands leads also to estimates of atmospheric temperatures.

Radiometric observations of the infrared radiation by means of thermal receivers such as the thermocouple provide the most powerful method for the determination of planetary temperatures. Our present scanty knowledge could be greatly advanced by performing such observations from space probes at close range from the planets.

Radio observations at centimeter wave lengths are of growing importance for the determination of the surface temperatures of planets with dense atmospheres, such as Venus, not penetrated by ordinary infrared radiation, or perhaps for determining electron densities in their ionospheres, if these should prove to be the origin of the observed radiation. At millimeter wave lengths atmospheric absorption bands, especially of water vapor, may be detected and at meter wave lengths non-thermal noise may help monitor atmospheric electrical storms. At still longer wave lengths, in the decameter range, bursts of non-

The Moon and Planets

thermal radio radiation, such as are observed from Jupiter, may help solve the problem of the internal constitution of the giant planets.

Only weak radar echos have yet been received from the planets, but successful radar probing of the lunar surface shows that the method holds great promise for future application to the planets, especially when the solid surface is hidden by clouds.

OBSERVING STATIONS

These various types of observations can be performed from a variety of stations, each with different advantages or disadvantages, and we should not encourage the still frequent error of considering that space exploration can only be performed from a vehicle in space. After all, astronomers have been quite successfully exploring space, near and far, for over two thousand years.

Ground-based observatories remain of fundamental importance for obvious reasons, and their permanent value for the progress of lunar and planetary exploration should not be underestimated. As Professor Harold C. Urey remarked recently, "We propose to spend enormous sums of money to fly apparatus away out there, yet ignore the handy little devices that lie on our doorstep. It's a very inconsistent attitude."

Airplanes are of little use for observations of the moon and planets because of vibration and other difficulties, but they are the cheapest and most readily available "space" stations to observe the planet earth from above. Such studies are essential not only for their intrinsic interest to meteorologists, but also as an invaluable guide for the interpretation of planetary observations.

Stratospheric balloons are destined to become of increasing importance to astronomical observations. Floating above 98 or 99 per cent of the atmosphere, they combine most of the advantages of true space stations with the lower cost and greater convenience of operation, including easy retrieval of information and recovery of equipment, associated with ground-based operations. High-resolution photographs unimpeded by atmospheric turbulence, sensitive spectral analyses free of troublesome atmospheric obstruction, and detailed thermal studies of the moon and planets will become possible through the systematic use of stratospheric observatories. Establishment of a national facility available to astronomers, meteorologists, geophysicists, and physicists is an urgent need for the orderly development of the space program.

Rockets have been successfully applied to solar studies where exposure times are short. They do not appear promising at the moment for the study of the moon and planets. In the future, it might become possible to launch rockets from space stations to perform experiments in the upper atmospheres of other planets.

Artificial satellites of the earth will be most valuable, indeed essential, for studies of the vacuum ultraviolet region of the spectrum. They will, of course, have also the advantages of balloons in regard to freedom of atmospheric interference. However, in assessing the relative merits of balloons and satellites for planetary research it must be kept in mind that satellites of the earth will not be significantly closer to the planets than ground stations, while the successful operation of a large telescope from a satellite for high-resolution studies is bound to be much more difficult than if it is attached to a stratospheric balloon.

Space probes will be the tool *par excellence* of direct planetary exploration. It is only through "close-up" photography from space probes that detailed mapping of the planets with resolution better than one mile will become possible. It is only through a close approach to a planet that its magnetic field can be measured, its radiation belts penetrated, and its ionosphere probed by radio sounding in any detail. Most of the observations that can be performed from the earth, from balloons, and from satellites could be repeated with much greater resolution—at the cost, of course, of much greater technical difficulties and, initially, much lower probability of success. An almost limitless list of experiments can be considered; detailed studies of priorities in relation to scientific significance, technological capabilities, and available payloads become of crucial importance for the planning of the space program.

TYPES OF EXPERIMENTS

The mere listing of potential experiments could cover many pages and would encompass essentially the whole range of human and technological knowledge, from mathematics and information theory to biology and space medicine. A rough classification according to the probable time sequence imposed by technological progress follows:

1. Interplanetary probes on one-way flights during the period of closest approach (fly-by or near-miss) can make brief, primarily passive, physical and mechanical observations of the planet and its environment. Such observations will include, for example, a search for magnetic fields and radiation belts; a search for unknown satellites; high resolution photography and television of the surface; measurements leading to improved determinations of planetary masses and diameters (especially of the shape of the moon elongated toward the earth); refined photometric, spectroscopic, and polarimetric studies of planetary surfaces and atmospheres, of sunset and twilight phenomena, of the luminescence of the atmospheres on the night side of Mars and Venus; scans of the infrared thermal emission on the day and night hemispheres; passive detection of the microwave thermal and non-thermal radio emissions, and active ionospheric radio sounding and radar probing. And these are only some of the most obvious possibilities.

2. Artificial satellites (orbiters) placed in a controllable closed orbit can make both passive and active physical and mechanical observations and experiments for significant periods of time before they escape on an earth-bound trajectory or burn up by atmospheric penetration. Such experiments would include more precise measurements of the mass and radius of the planet, of its dynamic and physical ellipticity; detailed optical mapping with resolution down to a few meters; finer photometric, spectroscopic, infrared, microwave, radar, and magnetic surveys; direct determinations of upper atmospheric densities and upper atmospheric absorption; studies of ionization phenomena and of upper atmospheric winds and turbulence.

3. Penetrating probes, carrying simple physical, chemical, and meteorological sampling apparatus through the atmosphere for

a short period of time prior to a (generally destructive) crash landing with loss of communication, can make direct measurements of characteristics including atmospheric pressure, density, and temperature as a function of altitude above the surface, of atmospheric transmission and scattering, and of twilight phenomena.

4. Non-destructive crash landing without communication loss (using retro-rockets or parachutes) of rugged physical, chemical, and geophysical apparatus at the surface of a planet or the moon should allow simple passive observations of mainly local conditions. Such observations would include measurements of the velocity of sound; penetrometer and hardness tests; passive and active seismic recordings; high speed, high resolution television of the impact area prior to landing; records of surface temperatures, and the like.

5. Once soft landings without communication loss can be achieved, simple physical, chemical, and biological sampling equipment can be used for passive and active observation and experimentation at or near the impact point. For instance, detailed spectral measurements of atmospheric absorption both optical and radio, of atmospheric surface temperatures and pressures, of the thermal radiation from the atmosphere during day and night, of atmospheric humidity and composition—in brief the basic elements of the local "climate"—could all be made. Studies of light emission at night to monitor lightning, aurora, airglow, of radio emission ("atmospherics"), of atmospheric electricity, and the like, could also be made. Accurate measurements of the surface magnetic field, of the acceleration of gravity, of surface radioactivity (especially for the determination of the uranium, thorium, potassium abundances on the moon, Mercury, Venus and Mars) will be of primary importance for the comparative geology—"planetology"—and cosmogony of the terrestrial planets. Spectroscopic analysis of atmospheric dust and surface rocks should be attempted to help determine the composition of the surface. Close-up television of the landscape and surrounding terrain, and microscopy of atmospheric dust and rocks at the landing site, will have also high priority in the first direct exploration of these new worlds.

The Moon and Planets

Listening for atmospheric noises with directional acoustic devices may also be of interest for Mars, especially if prior optical reconnaissance indicates the presence of forms of higher life. Methods to collect atmospheric particles and surface dust for possible biological sampling could be devised and adapted to microscopic examination.

6. Eventually, soft landings of more complex and mobile equipment for extended direct exploration and sampling of a wide area should precede the organization of manned expeditions, but this is clearly a still remote goal and steps 1 to 5 should suffice fully to occupy our generation! Here the danger of insufficient preparation, of hasty and wasteful space shots for the sake of shooting first should be recognized and, if possible, avoided.

NECESSARY EARTH-BASED STUDIES

Even before we seriously consider plans for direct planetary exploration, it is essential for the proper planning of the space program, and later for the proper interpretation of its results, that an extensive and intensive back-up program of earth-based studies be set up and organized on a commensurate scale. The most obvious needs are: in mathematics—orbit theory and flight paths, information theory and optimization of data transmission and processing; in astronomy and atrophysics—systematic planetary observations with ground-based and balloon-borne telescopes, teaching and training in physical planetology (one of the most urgent needs in this country), and development of special instrumentation; in physics—scattering theory and laboratory measurements, especially of the optical properties of a great variety of terrestrial samples, and improvement of more sensitive radiation sensors and detectors; in chemistry and photochemistry —experimental study of plausible planetary atmospheric and surface constituents under a variety of physical conditions, and determination of cross-sections; in meteorology—theory of planetary atmospheres and circulation; in geology and minerology —new analytical tools and techniques; in biology and exobiology—contamination problems; not to mention the vital need for technological advances in the field of miniaturization of rugged laboratory apparatus and communications equipment.

Basic scientific significance appears to be, in the long run, the main rational justification for a sustained space effort and the solid foundation for whatever practical ends may eventually accrue from it (and undoubtedly will, as the whole history of science shows). It is therefore essential that the space effort should be solidly anchored to a well-rounded program of scientific development on this earth.

In the fields of immediate interest for lunar and planetary exploration the most pressing needs are:

First, selective and continuing adequate support through research grants to the few active groups with previous experience in the subject; second, strong support, in particular through scholarships, for teaching at the graduate level leading to the formation of a new generation of "planetologists" with a solid education in astronomy, meteorology, geophysics, and allied sciences; third, substantial financial support for the development of telescopic and auxiliary instrumentation, including, for the present, telescopes specially designed and available full time for lunar and planetary work at established ground-based observatories in good locations, and, even more important for the near future, a massive and immediate effort to launch the powerful balloon facility for a national stratospheric observatory — a necessity repeatedly stressed by several conferences and study groups during the past three years.

SUGGESTED READING

JASTROW, R., (ed). *The Exploration of Space.* New York: Macmillan, 1960.

KUIPER, G. P., and MIDDLEHURST, B. *Planets and Satellites.* Chicago: University of Chicago Press, 1957.

MOORE, P. *The Planet Venus,* 3d ed. London: Faber & Faber, 1960.

SAGAN, C. "The Planet Venus," *Science,* 1961.

DE VAUCOULEURS, G. "Mars," *Scientific American,* **188** (1953) 65–73.

——— . *Physics of the Planet Mars,* London: Faber & Faber; New York: Macmillan, 1954.

———. "Planetary Observations from Space Probes and Orbiters," *Proceedings of Space Age Astronomy Symposium,* Pasadena, August, 1961 (in press).

———. *The Planet Mars.* 2d ed. New York: Macmillan, 1954.

PART THREE

National Space Programs

Introduction
United States Space Program
NASA and Space
Department of Defense Space Program
Space Programs of Other Nations

INTRODUCTION

HUGH ODISHAW

The fact that only two nations now have the capability of launching spacecraft tends to conceal the appreciable activities and breadth of interest in space on the part of many nations in addition to the United States and the Soviet Union. It also casts into shadow the likelihood that other nations will in time acquire space launching capability if they wish to do so. But we can expect the number of satellite-launching countries to increase: already a group of Western European nations has begun negotiations directed toward a multi-nation launching authority.

This section describes current national space activities throughout the world. It emphasizes the activity and structure within the United States not only because space is a major effort but because information is readily accessible. Work within the Soviet Union is also summarized, on the basis of published literature, but information on organization, structure, and related topics is not available. Finally, the space contributions of other nations are outlined.

To sense the nature of the world's effort in space at the present time, it is useful to separate space activity into four areas: (1) rocket systems, (2) man in space, (3) applications, and (4) research. It is also useful to bear in mind that space activities are not restricted solely to endeavors conducted aboard satellites and space probes. Balloon and sounding rocket studies, investigations at ground stations by classical techniques, and theoretical studies are just as pertinent, and they are relevant not only to progress in research and technology but to space flights. Thus intensified astronomical studies of the moon and planets can yield new knowledge of interest both to the research-minded astronomer and to the planner of a manned mission to the moon. The larger the body of data at hand about the lunar or Martian environment, the surer the planning for a mission to that body. If such data can be obtained from the earth's surface, so much the better because costs are lower and payload space is left open for tackling problems that require space systems.

In the four areas of space activity noted above, the following general characterizations can be made. Rocket systems of the kind needed to launch satellites, space probes, and planetary adventures represent enterprises that only the United States and the Soviet Union now support. The costs are such that within the next few decades it is unlikely that other nations, or even groups of nations, will enter the lists of manned ventures. It is likely that other nations, or groups of nations, will undertake satellite-launchings for research and application and small space probes for research. The field of sounding rockets is even more open: costs are not excessive; several nations have developed such rockets for research in the upper atmosphere, and more work along these lines is almost certain to be done. Somewhat the same remarks apply to balloon research and apply even more to research activities based on ground observations and on laboratory and theoretical investigations.

The quantity and quality of work under way throughout the world on space or space-related fields of investigation is appreciable. The larger nations do more, as might be expected, and among the larger nations those that have greater resources and perhaps longer traditions in research inevitably do the most. But generalizations are dangerous: the small state of New Zealand does more than some nations many times its size and, to cite a specific example, New Zealand contributed a larger program to the International Geophysical Year, on a per capita basis, than the United States, whose total effort was probably the largest of all nations.

The space efforts of the United States are extensive — approaching expenditures of five billions per year—and they probably will increase in magnitude for several years. The program is also a complex one, as suggested by the three following chapters which discuss its organization and the civilian and military roles.

The interests of the military services, which currently spend some one to two billion dollars each year on space or space-

Introduction

related activities, are several. First, work is under way on such satellite systems as Midas and Samos, for reconnaissance and early warning. Second, the services have interest in communication, weather, and geodetic satellites. Third, the development of space rocket systems goes on and, in particular, the Air Force is concerned with large-thrust, solid-propellant rockets. Fourth, there are military interests in the physical environment of space, and research is conducted by or under the auspices of the services on such topics as the physics and chemistry of the upper atmosphere and near space, cosmic rays, and solar-terrestrial relationships. The bulk of military activity falls within the purview of the Air Force, but both the Army and the Navy are active in areas pertinent to their missions.

The major effort of the United States in space has been assigned to the National Aeronautics and Space Administration (NASA). Its budget during the fiscal year was almost two billions, in 1963 it will be about four billions, and in 1964 perhaps five to six billions. The increase in NASA's budget reflects the increased national interest in space activities, which received its major impetus in the spring of 1961, when President Kennedy announced the manned lunar effort as a national goal and set its attainment within the present decade. This program alone, already under way as Project Apollo, will cost some 50 billions, over the course of the next seven to ten years, before its objectives are attained. At the present time more than half of the NASA budget is devoted to man in space. For an over-all, bird's eye view of NASA's activity, the total program may be broken into six categories: (1) man in space, 60 per cent of the budget; (2) unmanned scientific investigations of all kinds, 14 per cent; (3) space technology, a general area of support for all space activity, 14 per cent; (4) other supporting operations, largely tracking, 6 per cent; (5) space applications, 4 per cent; and (6) aeronautics, 2 per cent. This is an oversimplification: for example, while 2 per cent of the budget is assigned to aeronautics, 10 per cent of NASA's staff is devoted to this field; moreover, this does not reflect the present utility of such expensive facilities as wind tunnels, acquired in the past. Because any attempt to describe in brief the NASA enterprise represents an oversimplification, not much is lost by going a step further. By ignoring space technology and supporting operations as separate divisions, by distributing funds for these (totaling together 20 per cent) among the three space areas of effort, and by adjusting the aeronautics program to reflect something like the expenditures involved were it a new effort requiring the capital and back-up costs built up over many years, one can speculate that the current NASA allotments run about as follows: (1) man in space, 70 to 75 per cent; (2) space research, about 15 per cent; (3) space applications, about 5 per cent; and (4) aeronautics, about 5 per cent.

The magnitude of the nation's energies devoted to man in space has raised public questions. Some scientists as well as public leaders have argued that man in space has little value and that the large sums devoted to this effort could more wisely be spent in other ways—from cancer research to housing. Other scientists and national leaders have taken quite the opposite position. Their arguments assume one or more of the following forms. Coin in one realm cannot be exchanged for coin in another: if the manned space effort were abolished, there is no evidence that the "saved" funds would be available for other purposes. The effort is justifiable because adventure is a legitimate human drive, because there is a competitive situation between the world's two chief powers, and because the undertaking has many by-products which in themselves warrant the magnitude of the program. These by-products include the development of a new technology and industry whose impacts may well reach far into the total economy, the development of space systems for man that will have utility in other space programs, and the ultimate scientific interest in the manned, detailed exploration of the moon and the nearer planets.

Put into slightly different words, one could argue that men and nations tend to be responsive to great challenges. Men and nations have competed for laurels and leadership, and competition *per se* is not necessarily undesirable provided that co-operation among nations ameliorates a strictly political race. Man is someday bound to get to the moon and the nearer planets: if so, the question is one of when and not of why. The question

Introduction

whether the costs of reaching the moon would be less if the program were stretched over two or three decades instead of one cannot be answered. It is likely that theoretical savings in a more leisurely space system development would be at least offset by indirect costs associated with capital investments and overhead.

One independent group of scientists, the Space Science Board of the National Academy of Sciences, came to the following conclusion (quoted in part) after meditating on the man-in-space question for some three years, in a statement prepared early in February of 1961:

> The Board concluded that it is not now possible to decide whether man will be able to accompany early expeditions to the Moon and planets. Many intermediate problems remain to be solved. However, the Board strongly emphasized that planning for scientific exploration of the Moon and planets must at once be developed on the premise that man will be included. Failure to adopt and develop our national program upon this premise will inevitably prevent man's inclusion, and every effort should be made to establish the feasibility of manned space flight at the earliest opportunity.
>
> From a scientific standpoint, there seems little room for dissent that man's participation in the exploration of the Moon and planets will be essential, if and when it becomes technologically feasible to include him. Man can contribute critical elements of scientific judgment and discrimination in conducting the scientific exploration of these bodies which can never be fully supplied by his instruments, however complex and sophisticated they may become. Thus, carefully planned and executed manned scientific expeditions will inevitably be the more fruitful. Moreover, the very technical problems of control at very great distances, involving substantial time delays in command signal reception, may make perfection of planetary experiments impossible without manned controls on the vehicles.
>
> There is also another aspect of planning this country's program for scientific exploration of the Moon and planets which is not widely appreciated. In the Board's view, the scale of effort and the spacecraft size and complexity required for manned scientific exploration of these bodies is unlikely to be greatly different from that required to carry out the program by instruments alone. In broad terms, the primary scientific goals of this program are immense: a better understanding of the origins of the solar system and the universe, the investigation of the existence of life on other planets

and, potentially, an understanding of the origin of life itself. In terms of conducting this program a great variety of very intricate instruments (including large amounts of auxiliary equipment, such as high-powered transmitters, long-lived power supplies, electronics for remote control of instruments and, at least, partial data processing) will be required. It seems obvious that the ultimate investigations will involve spacecraft whether manned or unmanned, ranging to the order of hundreds of tons so that the scale of the vehicle program in either case will differ little in its magnitude.

The Board strongly urges official adoption and public announcement of the foregoing policy and concepts by the United States government. Furthermore, while the Board has here stressed the importance of this policy as a *scientific* goal, it is not unaware of the great importance of other factors associated with a United States man in space program. One of these factors is, of course, the sense of national leadership emergent from bold and imaginative United States space activity. Second, the members of the Board as individuals regard man's exploration of the Moon and planets as potentially the greatest inspirational venture of this century and one in which the entire world can share; inherent here are great and fundamental philosophical and spiritual values which find a response in man's questing spirit and his intellectual self-realization. Elaboration of these factors is not the purpose of this document. Nevertheless, the members of the Board fully recognize their parallel importance with the scientific goals and believe that they should not be neglected in seeking public appreciation and acceptance of the program.

UNITED STATES SPACE PROGRAM

WILLIS H. SHAPLEY

This review of the present organization of the United States government's programs for the exploration and utilization of outer space will outline the organizational pattern and attempt to show how space activities are woven into the general fabric of government. The actual programs of the two principal agencies mainly involved in the space program, the National Aeronautics and Space Administration and the Department of Defense, are fairly well known, and are discussed in other chapters. Activities of other agencies, such as the Weather Bureau (meteorological satellites) and the Atomic Energy Commission (nuclear energy for space vehicles) are also discussed elsewhere.

Some of the organizational relationships among these agencies, and the reasons underlying the division of responsibility and working arrangements, are less widely known. A certain mystery sometimes surrounds the roles of such agencies as the National Aeronautics and Space Council and the Bureau of the Budget. The scope and functions of the several committees of Congress which are involved with the space program are not always understood. The fact is that the government does not, and cannot, have a simple, clear-cut, and unified operation to carry forward its interests in space. The reasons for this lie partly in the complexity of modern government and partly in the nature of space activities.

The problems of government organization were of staggering proportions long before there were large space programs to contend with. In particular, the organization of the defense establishment—internally and in relation to similar activities of civilian agencies—and of large-scale scientific and technical activities have always presented problems not amenable to clear-cut or simple solutions. In these and in other areas, whenever a particular field such as basic research, aviation, guided missiles, electronics, medical services, or weather activities, has attracted special interest, there have been proposals and pressures to bring together into a single organization all the govern-

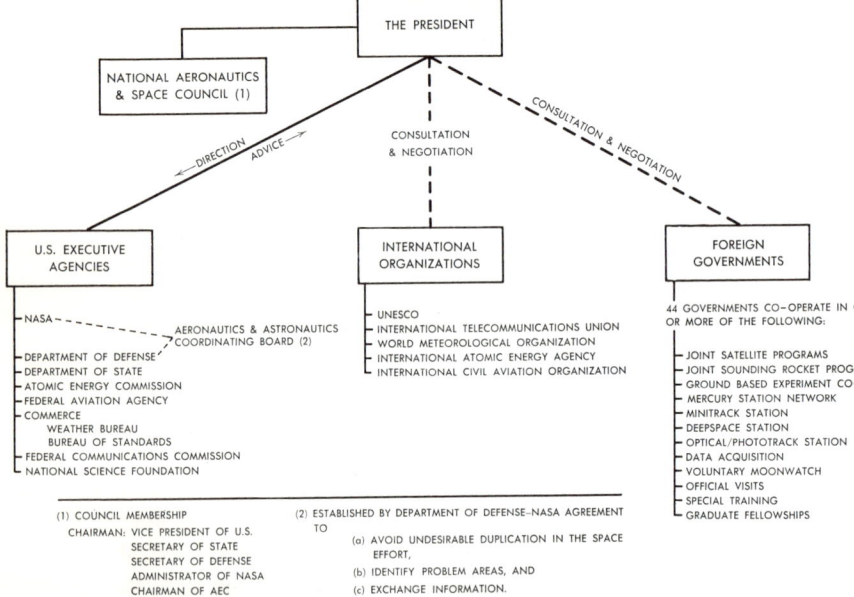

Fig. 1—Governmental and inter-governmental organizations in the U.S. space programs. Chart prepared by the United States Senate Subcommittee on Reorganization and International Organizations.

ment's activities in that field. This rarely proves to be practical. Each of the functions of government is so interrelated with many others that creation of a simple organization along one dimension of interest will generally cause serious dislocations along others. For this reason, many bold and superficially attractive organizational departures, such as establishment of a Department of Science or revival of the World War II Office of Scientific Research and Development, have been considered and rejected over the years as not representing practical methods of dealing with the total problem. At any given time, governmental structure represents a set of compromises among many (often incompatible) dimensions of interest. The nature of the compromises is determined, of course, by historical and political factors as much as by logic.

The sudden recognition of space activities as a major national interest led almost at once to the idea that a new and separate government agency was required. In this case the need for organizing an agency of the government along a new dimension of interest was convincing. A major effort in a new field was required, and it seemed clear that it could not be undertaken by existing military organizations without controverting or undermining the national posture stressing the peaceful uses of outer space and international scientific co-operation. But creation of a single governmental agency for a seemingly clear-cut purpose raised complex and difficult questions with respect to its scope and relationships to the other necessary functions and agencies of the government, particularly those of the Department of Defense.

From the start it was clear that there were at least some military aspects of space which were of great possible significance to the United States, such as, for example, reconnaissance and satellites for early warning of missile attack. Conceivably, the civilian space agency might have been given the function of developing and perhaps producing military space systems, comparable to the Atomic Energy Commission's role in the development and manufacture of atomic weapons. Whatever its theoretical merits, however, this alternative was never a practical possibility. Unlike atomic energy, space technology has not developed as a distinct and novel undertaking under centralized direction. Its foundations were in the military missile programs which, for over a decade, were spread among many separate organizations in the Department of Defense. The principal technical fields involved—rocket propulsion, guidance and control, electronics, structures, and materials—were and still are more closely related to their counterparts in the military missile programs than they are to each other, or to a newly defined area of space technology.

These inescapable relationships between the space and missile programs, ranging from clear-cut cases like the use of the same vehicles, components, and launching facilities, to hard-to-define areas where the same supporting research and development contributes to both programs, have made the idea of consolidating all space activities in a single civilian agency an academic one, and are at the root of many of the organizational problems that still exist.

CONDUCT OF THE SPACE PROGRAMS

There are five principal agencies responsible for conducting the space programs of the government: the National Aeronautics and Space Administration, the Department of Defense, the Atomic Energy Commission, the Weather Bureau, and the National Science Foundation. The present pattern of organization has evolved gradually since the period of initial reaction to Sputnik I in the fall of 1957. At first space activities, except for the Vanguard project under way in the Navy and sponsored jointly by the National Science Foundation and the Department of Defense, were centralized in the newly established Advanced Research Projects Agency (ARPA) in the Department of Defense. In the summer of 1958, the National Aeronautics and Space Administration (NASA) was established as an independent civilian agency, generally responsible for non-military space programs.

NASA absorbed the functions of the former National Advisory Committee for Aeronautics and its laboratories, which were already engaged in research related to space vehicles as well as to aircraft and missiles. Vanguard and certain projects that had been initiated by ARPA and the Air Force were transferred from the Defense Department to NASA immediately. Later the Army's Jet Propulsion Laboratory, operated by contract with the California Institute of Technology, and virtually the entire Army Ballistic Missile Agency at Huntsville, Alabama, were also transferred to NASA.

The space projects remaining in the Department of Defense continued for a time to be under the direction of ARPA, although most of them were carried out by the military services serving as ARPA's agents. In the fall of 1959 the Air Force was given responsibility for most of the defense space programs, a decision which was reaffirmed in March of 1961. The Air Force is presently responsible for all major military space programs, except for the navigation satellite program which is assigned to the Navy, and communications satellite ground stations which are assigned to the Army. Except for the satellite aspects of the Vela-Hotel program for the detection of nuclear explosions in space, ARPA is no longer responsible for space projects.

The general theory of the division of responsibility for space projects between NASA and the Department of Defense is stated in the National Aeronautics and Space Act of 1958. NASA is responsible for all space activities,

. . . except that activities peculiar to or primarily associated with the development of weapons systems, military operations, or the defense of the United States (including the research and development necessary to make effective provision for the defense of the United States) shall be the responsibility of, and shall be directed by, the Department of Defense; and that determination as to which such agency has responsibility for and direction of any such activity shall be made by the President. . . .

The Department of Defense has general responsibility for matters related to national defense including such uses of space. NASA is organized to deal with space; the Department of Defense is organized to deal with defense. The area of overlap is hard to define and impossible to divide on a logical, mutually exclusive basis. The role and responsibilities of each agency must largely be determined by specific decisions on specific projects. The existence of apparent duplication — substantial areas in which both agencies are active—must be accepted as inherent in the situation.

Under the present division of responsibilities, NASA is generally responsible for space science, for exploration of space, both unmanned and manned, and for the development of nonmilitary applications of space vehicles. NASA is also responsible for the development of space vehicles and their components and for conducting a broad program of supporting research and development (including aeronautical research related to both aircraft and missiles), but these responsibilities are shared in varying degrees and in various ways with the Department of Defense.

As a result of specific decisions, NASA is now responsible for the development of most of the new space vehicles in the current national program. These include the Scout, the Thor-Delta, the Atlas-Centaur, the Saturn, and the Nova and its liquid propellant booster. The Department of Defense is currently responsible for developing the Thor-Agena and Atlas-Agena combinations, an unnamed possible alternative to the Atlas-Centaur, and a solid propellant booster for possible use with the Nova vehicle or other future systems. Each agency makes use of vehicles developed by the other. NASA uses the Thor-Agena and Atlas-Agena in the meteorological satellite program. The Air Force is using a version of the Scout in certain experiments, and the Advent military communications satellite program, recently canceled, was to have used the Atlas-Centaur.

NASA is now primarily responsible for the development of manned space flight. Full responsibility for the Mercury program was transferred to NASA from ARPA at the time NASA was established. The huge manned lunar landing program announced by the President on May 25, 1961, is the responsibility of NASA, except for the development of large solid propellant booster engines and technology, which was assigned to the Air Force. The Dynasoar project for developing a rocket-boosted manned glider, which may eventually have orbital capabilities, is being conducted by the Air Force with some technical assistance from NASA. The X-15 research airplane has been a joint NASA–Air Force project, with some Navy participation, and is now in phases for which NASA has primary responsibility.

In the many different fields of supporting research and development, both NASA and Defense are deeply involved. For reasons that have been mentioned, no clear-cut divisions of responsibility exist or are possible, and it is difficult to describe the situation in general terms. In addition to the supporting research and development directly generated by specific space programs, the new interest in and increased support for space activities appear to have caused a noticeable shift in emphasis (in both government laboratories and industrial organizations) to research and development relevant to the problems of space vehicles and their planned and potential applications.

There are many organizational interactions between NASA and the Department of Defense in space matters besides those involved in the division of responsibilities for particular programs. The most important is the joint use of launching and range facilities. The Department of Defense operates, through the Air Force and its contractors, the tremendous complex known as the Atlantic Missile Range, which extends from the launching area at Cape Canaveral to range ships which may be stationed as far away as the Indian Ocean. As is well known, these facilities are used for most NASA major space launchings and vehicle development flight tests, as well as for some Defense space programs and, of course, for their original military purpose of testing guided missiles. A similar pattern is developing at the Pacific Missile Range (operated by the Department of Defense through the Navy) with respect to both Defense Department and NASA space launchings requiring polar or high inclination orbits which are not feasible from Cape Canaveral. To some extent, the same tracking and data acquisition sites and facilities are used by both NASA and the Defense Department. However, the degree of joint utilization is limited by the fact that for many of the projects of both agencies special requirements of equipment or geographical location are essential.

Two different types of NASA–Department of Defense relationships are significant enough to mention. One is the participation of Department of Defense laboratories, and sometimes their contractors, in developing scientific payloads for satellites or other space vehicles. These laboratories provide experiments for NASA sounding rockets, scientific satellites, and space probes on substantially the same basis as the scientific community at large and the NASA's own research and space flight centers. In addition, some experiments from all these sources are flown "piggy-back" in both military and NASA space launchings primarily directed at vehicle development or other objectives. Another NASA–Department of Defense interrelationship is the "cross-servicing" that exists in the actual conduct of the programs. For example, all Thor and Atlas boosters are procured for both NASA and the Defense Department by the ballistic missile organization of the Air Force. Special management arrangements have been worked out for most of the major projects of each agency in order to utilize existing facilities, organizations, and capabilities in matters ranging from technical direction and the conduct of tests to contracting, purchasing, and auditing.

Organization for space applications deserves special comment. It is not only a matter of assignment of responsibility for development of each system to either NASA or the Department of Defense, but there is the additional question of what agency of the government should continue to have responsibility for operating the system as a going concern. For purely military applications, the answer is simple enough. For applications of broader interest, specific decisions have to be made. Military interest in navigational satellite systems is predominant enough to warrant assignment of both the development and the operation of such a system to the Navy. The Transit project is being carried forward on this basis, and will provide a system available to both civilian and military users.

The opposite conclusion was reached for meteorological satellites. NASA is conducting the Tiros and Nimbus projects to develop a primarily civilian system to serve both civilian and military requirements. An organizational compromise was necessary between the Weather Bureau's general responsibilities for meteorology and NASA's general responsibilities for space. The decision has been that NASA, in consultation with the Weather Bureau, is responsible for the development of a meteorological satellite system, and that the Weather Bureau will be responsible for its operation. To avoid putting the Weather Bureau itself directly into the business of procuring and launching space vehicles, NASA will perform these activities for the operational system, acting as an agent for the Weather Bureau.

In the communications satellite field, the details of the ultimate pattern are not yet clear. NASA has the responsibility for the government's activities in developing a system for commercial use, and in testing components for such a system developed by private concerns. The policy and expectation is that the actual commercial system will be established and operated by a private corporation for profit under appropriate government regulation, with the satellite launchings performed by NASA on a cost basis.

The responsibilities of many government agencies besides NASA impinge on the communications satellite program. The Federal Communications Commission's regulatory functions are directly involved. Frequency allocation problems, until recently the responsibility of the Office of Civil Defense and Mobilization, are now assigned to its successor agency, the Office of Emergency Planning. International aspects which are crucial to the success of a world-wide system mean that the State Department is concerned. Possible antitrust situations that might develop bring in the Department of Justice. The Department of Defense is also interested, as one of the principal government users of such a system when it becomes available for general use.

The Department of Defense is also working on purely military communications satellite systems, on the joint NASA–Defense Syncom project, and on other experiments related to possible space communications systems, such as the "needles" project, West Ford, which has recently received considerable criticism from some of the scientific community because it, or successors, may possibly interfere with astronomy. The first attempt at orbiting the needles was a failure; the needles did not separate.

The principal role of the Atomic Energy Commission (AEC) in the space program is in developing nuclear reactors for use in space vehicles. The well-publicized Rover project, a nuclear rocket propulsion system for space vehicles, and the Space Nuclear Auxiliary Power systems (SNAP) program for developing a nuclear source of electric power for space uses, are the two main lines of endeavor. These programs, described elsewhere in this section, present another organizational intersection of two dimensions of interest. In the case of nuclear rocket development, the answer has been to establish a joint AEC-NASA project office to manage the program. It would not make sense either for NASA to enter the AEC's specialized field of reactor development or for the AEC to become a second space agency developing and testing only nuclear rockets. The interrelations are not so close in the development of nuclear sources of electric power. Devices suitable for use in space impose many specialized developmental requirements but may have applications in other fields. The AEC currently carries on its work in this field without special organizational arrangements, but in close coordination with NASA and other interested agencies.

The National Science Foundation (NSF) has not participated in space activities in a major way since its early support of the Vanguard project in connection with the International Geophysical Year (IGY). Government participation in the IGY space effort stemmed largely from recommendations and plans of the National Academy of Sciences, whose IGY committee was responsible for United States contributions. The high altitude sounding rocket program, jointly supported by NSF and other agencies during the IGY, has been continued primarily by NASA, as a part of its space science program. NSF and NASA jointly support the National Academy of Sciences' Space Science Board, a domestic advisory body which also promotes international space co-operation through the International Committee on Space Research (COSPAR). The principal space project supported currently by NSF is the development of a space telescope which is a part of the program of the Kitt Peak National Observatory. NSF also gives substantial support to astronomy, particularly radioastronomy, as well as to other fields which could be considered, in a broad sense, space activities.

CO-ORDINATION OF SPACE PROGRAMS

There are several agencies in the executive branch which are concerned with the co-ordination and review of the space programs. Four deserve mention: the Aeronautics and Astronautics Co-ordinating Board, the National Aeronautics and Space Council, the Office of Science and Technology, and the Bureau of the Budget.

The Aeronautics and Astronautics Co-ordinating Board (AACB) coordinates the many interactions between NASA and the Defense Department in space activities discussed above. The AACB was established in September, 1960, by joint action of the Secretary of Defense and the administrator of NASA after experience had shown that the former Civilian-Military Liaison Committee provided in the Aeronautics and Space Act of 1958 was not an effective mechanism. The deputy administrator of NASA and the director of Defense Research and Engineering serve as co-chairmen of the AACB. The other eight members of the board consist of four senior officials each from NASA and the Department of Defense; all are directly concerned with the management of space (or aeronautics) programs in their agencies. There are six panels covering fields such as launching vehicles, manned space flight, and space flight ground environment, each presided over by a NASA and a Defense Department member of the AACB. The membership is composed of senior specialists in the particular field from each agency. The board and its panels as such do not have any power of decision or action, but their members, by virtue of their positions in their own agencies, generally do have the necessary authority to take appropriate action. Any disagreements can usually be resolved by the co-chairmen of the AACB, or, if necessary, by the Secretary of Defense and the administrator of NASA. In its relatively short period of existence, the AACB has shown itself to be a reasonably effective mechanism for co-ordination in an extremely complicated situation.

The National Aeronautics and Space Council (NASC) is a unique mechanism for policy planning and co-ordination at the highest levels. As originally constituted in the National Aeronautics and Space Act of 1958, the council consisted of the President, who was to preside, and the Secretary of State, the Secretary of Defense, the administrator of NASA, the chairman of AEC, another member from a government agency, and three members from outside the government—forming a cabinet-like organization similar in some ways to the National Security Council. After about a year and a half, when the main outlines of the nation's space programs as then conceived had been set, the Eisenhower administration concluded that the NASC had outlived its usefulness. It was disbanded and the repeal of its statutory authorization was recommended.

The Kennedy administration took a different view. One of the first acts of the President after his election was to announce that the NASC would be revitalized under the leadership of the Vice-President. Legislation reconstituting the NASC was passed in April, 1961, making the Vice-President the chairman, eliminating the private members, and clarifying the function of the council, "to advise and assist the President, as he may request, with respect to the performance of functions in the aeronautics and space fields." The reconstituted NASC was provided for the first time with a full-time executive secretary, Edward C. Welsh, and a small full-time staff has now been built up. In its short history to date, the new NASC has taken an active part in successive reorientations of the space programs during 1961. Another significant contribution has been its development of a national policy with respect to communications satellites. It should be emphasized that neither the council, its executive secretary, nor its chairman has directive authority over the space programs of the United States government. Their function is to consider and to advise, as requested by the President. The chain of command and responsibility for space programs runs directly from the President to the administrator of NASA and the Secretary of Defense, and to the heads of any other executive agencies that may be involved.

The newly named Office of Science and Technology, until recently a part of the White House, is the highest focus for scientific and technical advice in the government. Its director, now Jerome B. Wiesner, and his staff, provide a source of scientific and technical advice not only to the President himself but to the other constituents of the executive office of the President, such as the Bureau of the Budget, the National Security Council, the National Aeronautics and Space Council, and to all the departments and agencies of government concerned with science or technology. The director also serves as chairman of the President's Science Advisory Committee, a permanent committee of outstanding scientists for general scientific and technical advice, and, as circumstances require, convenes standing or ad hoc committees for specialized advice in particular fields or on specific problems.

Under laws and Executive orders, some going back as far as 1921, the Bureau of the Budget is responsible for assisting the President in the discharge of his budgetary, management, and other executive responsibilities. The director of the Bureau of the Budget is one of the principal assistants to the President, and the bureau is a major component of the executive office of the President. The professional staff of the bureau consists of three hundred career civil servants, and provides the President with a central staff that is well informed on the operations and programs of all government agencies and the many problems the government faces. In space programs, the Bureau of the Budget performs the same functions as it does in all other areas of government activity. Hence a short explanation of its general functions will make it easier to understand what the bureau does in the specific field of space.

The Bureau of the Budget reviews the budget estimates and the current and proposed programs of all agencies of the executive branch of the government. This review not only represents an independent appraisal of the estimates and programs of the agencies, but also takes into account broader considerations such as interagency relationships and co-ordination, the relative priority of the various activities of the government, future costs of agency proposals, and the over-all fiscal position and policies of the government. The views and recommendations of the Bureau of the Budget together with those of the agencies are considered by the President in arriving at his annual budgetary recommendations to Congress. After appropriations are made by Congress, funds are apportioned to the agencies by the Bureau of the Budget, as required by law, in accordance with the policies of the President. The bureau monitors the programs and the financial operations of the agencies on a continuous basis throughout the year, to achieve greater economy in government operations and to identify for timely resolution problems and issues as they arise.

The Bureau of the Budget has specific responsibilities for the review and clearance of legislation proposed by government agencies, to assure that the proposals are in accord with the program of the President. It also has responsibilities for improving the organization and management of the government, including assisting in the establishment of new agencies and preparing and implementing reorganization plans. As a somewhat more specialized function, the bureau has the specific responsibility for co-ordinating statistical programs of the government and approving the forms used by government agencies to collect information from the public.

In discharging its normal functions, the bureau played a major role in the establishment and organization of the space programs. The original version of the National Aeronautics and Space Act of 1958 was drafted in the Bureau of the Budget, with the co-operation of other agencies, and submitted to Congress on behalf of the President by the director of the Bureau of the Budget. The reorganization actions transferring various space activities from the Defense Department to NASA were worked out by the bureau with the agencies involved. The Bureau of the Budget also necessarily played a major part in the decisions on the scope and nature of the space programs, and in each of the successive changes and reorientations that have taken place in both the Eisenhower and the Kennedy administrations. It was the bureau's function to assemble and review the various proposals and possibilities, and to present the major alternatives for consideration by the President, his top advisers, and the heads of the agencies directly concerned. This information formed the basis of the final decisions by the President. Long-range budgetary implications were especially significant in these considerations because of the commitment to tremendous future expenditures entailed by many of the possible space programs under consideration. For this reason, the bureau devoted special efforts to working out, with representatives of NASA and Defense, the best possible projections of future costs of each major program under consideration.

Within the Bureau of the Budget, the staff concerned with the space programs serves in the Military Division, which is responsible for carrying out the bureau's functions with respect to the Department of Defense, the AEC, and NASA. This means that the examiners dealing with civil and military space programs and with the related guided missile, military research and development, and atomic energy programs are all members of the same small group and can co-ordinate the interrelated problems of each area. In dealing with the space programs, as in other matters, the director and staff of the Bureau of the Budget work closely both with the agencies involved and with the other parts of the executive office. On space matters, close working relations have been established with both the NASC and the Office of Science and Technology.

CONGRESSIONAL ORGANIZATION

The two houses of Congress are organized to deal with space matters in much the same pattern as for most other fields of government activity. This involves two different kinds of committees—one for legislative matters and one for appropriations—in both the House of Representatives and the Senate. The reasons for separate committees in each House to deal with legislation and appropriations in the same field are deeply rooted in the history and traditions of Congress. The underlying theory is that the legislative committees are responsible for legislation authorizing the various activities of government, whereas the appropriations committees exercise the congressional prerogative of controlling the purse, and through it control the level, and indirectly the content, of the programs.

In addition to the division between the legislative and appropriative functions, congressional organization in the space field is further complicated by the fact that the basic pattern of committee organization in both Houses—for both legislation and appropriations—generally follows the existing organizational pattern in the executive branch, with each committee having jurisdiction over specified agencies or traditional fields of activity. The introduction of space activities as a new dimension of interest created much the same kinds of organizational complications in Congress as in the executive branch.

When the importance of space programs was inescapably

thrust upon the national consciousness by the Russian sputniks in the fall of 1957, both houses of Congress responded by appointing special "blue ribbon" committees. The majority leader in each House personally assumed chairmanship—Lyndon B. Johnson in the Senate and John W. McCormack in the House. The committees held numerous hearings on space and related subjects, culminating in the enactment by Congress in July of 1958 of the National Aeronautics and Space Act of 1958, a modified version of the President's proposal for establishing a civilian space agency. Shortly thereafter the special committees in each House were replaced by permanent standing committees with essentially the same membership.

In the Senate, the permanent committee is called the Senate Committee on Aeronautical and Space Sciences. Its legislative jurisdiction is limited to matters affecting NASA and other nonmilitary aspects of aeronautics and space. Space matters "peculiar to or primarily associated with the development of weapons systems or military operations" continue to come under the jurisdiction of the Senate Armed Services Committee, although the Committee on Aeronautical and Space Sciences is authorized "to survey and review, and to prepare studies and reports" upon such activities.

The permanent committee in the House has a broader scope. Named the House Committee on Science and Astronautics, its charter covers not only astronautical research and development, space exploration, and matters relating to NASA and NASC, but scientific research and development generally, scholarships in science, and matters relating to the Bureau of Standards and the National Science Foundation. This broad charter, however, is not interpreted as curtailing in any way the scope of the House Armed Services Committee with respect to military space and military research and development matters. While the new committees in both the House and the Senate are interested in space applications of atomic energy, the Joint Committee on Atomic Energy continues to have primary jurisdiction in all matters involving atomic energy and the AEC.

On the appropriations side, there is, strictly speaking, only one committee on appropriations in the House and one in the Senate.

However, both of these are divided into, and perform most of their functions through, a large number of subcommittees, each with jurisdiction over the appropriations of specific agencies of the government. In each House the Independent Offices subcommittee deals with the appropriations for NASA, whereas the subcommittee on Department of Defense appropriations considers the estimates for the military space programs together with those for all other military functions of the Department of Defense. Still another subcommittee in each house deals with the appropriations of the AEC. On the House side, each of the subcommittees tends to be a powerful separate entity, whereas on the Senate side there is overlapping membership among the appropriations subcommittees, and between the appropriations committee and the various legislative committees.

Congressional interest in space does not stop with those legislative and appropriations committees that have direct jurisdiction. Special investigating committees in each house concern themselves with space matters from time to time. Other standing committees have special interests, such as the concern of the committees dealing with the civil service with salaries paid to government scientists. The total effect is that the space programs receive their full share of congressional attention. The top officials of NASA and the Defense Department regularly appear before from four to six different committees to present and explain their programs, in varying detail. Congress requires the most formal procedure in the case of NASA. To secure its appropriations each year, NASA must first present and justify its program and estimates in detail to the House and Senate legislative committees and secure the enactment by Congress of legislation specifically authorizing the appropriations to be made. Then the same detailed justifications are presented to the House and Senate appropriations subcommittees, after which the actual appropriations are finally enacted by Congress.

Thus in both the legislative and executive branches we see that, while a few new organizations have been created, the new field of space activities has largely been absorbed into the regular structure and processes of government.

NASA AND SPACE

HOMER E. NEWELL

The mandate given to NASA in the National Aeronautics and Space Act of 1958 calls for a vigorous effort of science in aeronautics and space exploration. The act declares that activities in space should be devoted to peaceful purposes for the benefit of all mankind. It lists objectives for space activities in the United States, paraphrased as follows:

Expansion of knowledge of atmospheric and space science; improvement of aeronautical and space vehicles; development and operation of space vehicles; study of potential benefits to be gained for mankind through space activities; preservation of United States leadership in aeronautical and space science and technology, and in the application thereof to peaceful activities; interchange of information between civilian and national defense agencies; co-operation with other nations in aeronautical and space activities and in peaceful application of the results; effective utilization of the scientific and engineering resources of the United States in achieving these goals.

The first tasks facing the newly created NASA were to develop a sound program and to create an organization to carry it out. Fortunately the new agency had the existing National Advisory Committee for Aeronautics organization upon which to build. To this were added scientists and engineers of the Vanguard satellite team from the Naval Research Laboratory, the Jet Propulsion Laboratory, and later the Development Operations Division of the Army Ballistic Missile Agency at Huntsville, Alabama. From the outset, however, it was NASA philosophy not to develop an organization to carry out its entire program itself, but rather to rely heavily upon the national industrial and research resources. The effect of this philosophy is reflected in the present NASA program, in which more than 75 per cent of its budget goes directly into outside contracts, leaving less than 25 per cent to be expended on its own projects.

While recognizing that competition with the Soviet Union is a real factor in spurring United States space activities, the basic philosophy of NASA was to develop a space program that the country could undertake on its own merits. It was decided that the balanced program required vigorous efforts in aeronautics, vehicle development, application of space techniques and knowledge to practical uses, space science manned exploration of space, and supporting research and technology.

The newly formed NASA could rely on the Vanguard and rocket vehicles of the Department of Defense for initiating its space program. However, the capabilities of these vehicles were far short of those required to accomplish many of NASA's long-range objectives. It was necessary to undertake a program of vehicle development which could provide a steady growth toward the large payload capabilities required in many space applications and in the exploration of the moon and planets. It was desirable that this program should make use of the technology and vehicles from the missile program without interfering with defense needs, that it should serve all scientific needs, and that there should be a relatively small number of different vehicles in the program, consistent with the requirements of the flight program. Increased use of a small number of different vehicles enhances the reliability of each vehicle. Following this philosophy, NASA and the Department of Defense jointly worked out a national program for the development of United States space vehicles and are continuing to co-operate on it.

Several applications of space techniques to practical uses were apparent at the start of the program, but it was clear that considerable research was first required to lay the groundwork for the final practical applications. For this reason it was decided to press vigorously for a program of research directed toward determining the potentialities of a weather satellite and of passive communications reflector satellites. Following the initial success of the Echo passive satellite program, plans were extended to include research into the potentialities of active communications satellites.

The necessity for space investigation is a major justification for NASA's vehicle development program. Exploration and scientific investigation of space is also the principal justification for the development of manned space flight. The knowledge and techniques derived from space science research will also provide the basis for future practical applications. NASA started its space science program by building on the rocket and satellite research of the International Geophysical Year and the rocket sounding research of the previous decade.

Virtually all the IGY scientists who conducted rocket or satellite experiments during the IGY, and many more, have been drawn into the NASA program. It has been important from the beginning to draw heavily upon the national scientific community for the experimenters in NASA's space science program. It is expected, and intended, that the majority of the experiments and experimenters in the NASA satellites and space probes will come from outside of NASA. Counting past, present and definitely planned missions 60 per cent of the experimenters are from outside NASA, 40 per cent from inside. With the appearance of the large observatory satellites the ratio of outside to NASA experimenters should increase markedly. In the sounding rocket area, NASA has brought a number of new agencies into the program since the IGY. At the present time about 45 per cent of the space flight program budget (itself about a third of the total budget) is devoted to space sciences.

Finally, NASA philosophy calls for a strong program of supporting research and technological development to back up its aeronautics, vehicle development, applications, and space science programs.

Until recently the organization of NASA was as shown in Figure 1. The Office of Launch Vehicle Programs had the responsibility for vehicle development and for procuring and launching rockets and space vehicles required for the NASA space flight programs. The Office of Space Flight Programs had the responsibility for space sciences, applications, and manned space flight. Aeronautics, research on materials, propulsion and power supplies, came under the Office of Advanced Research Programs. With the recent announcement by President Kennedy

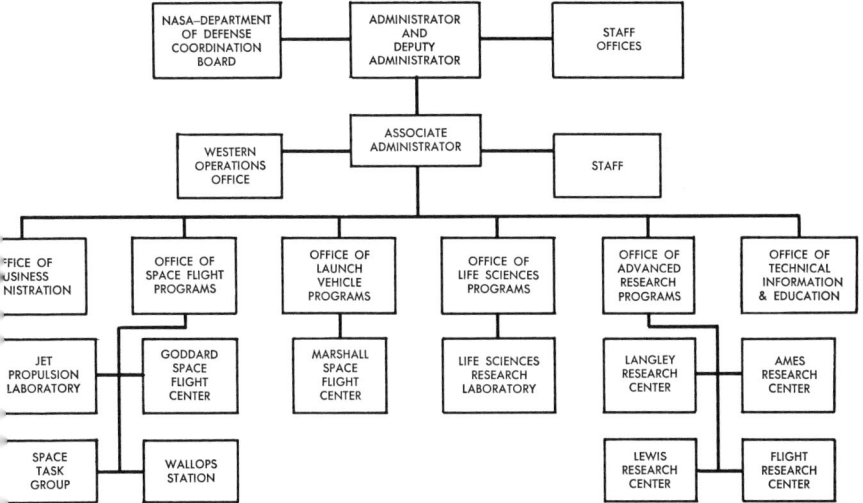

Fig. 1—Organization of the National Aeronautics and Space Administration.

that the United States will undertake to land men on the moon within the next decade, it has become necessary to revise this organization. In the new set-up the responsibility for manned flights and for space sciences are separated. The new organization places emphasis on the following four technical areas: advanced research and technology, manned flight, space sciences, and practical applications.

NASA SPACE PROGRAM

NASA has sought to develop a balanced program. For this reason the spectrum of activities is wide, ranging from aeronautics to manned exploration of space. A brief summary of these activities includes the following: In aeronautics NASA is continuing the work of the National Advisory Committee for Aeronautics. Some of the previous research has, however, been dropped or curtailed and new programs added, reflecting the needs and demands of commercial and military aviation. An important area of research is that of supersonic flight. NASA is conducting the X-15 program in co-operation with the Air Force, the Navy, and industry, using the X-15 for high speed-high altitude flight, aerodynamic, and related research. NASA is co-operating with the Air Force and industry in the Dynasoar project, which may produce valuable data on lifting re-entry and the problems of flight at speeds up to orbital velocities. NASA is also doing research and supporting development on vertical takeoff and landing aircraft, and on steep takeoff and landing aircraft.

NASA capabilities in space rest on the vehicles that are available. Earlier flights used the Vanguard, Atlas-Able, Jupiter C, Juno II, Redstone, and Thor-Able. These have now been retired. The present program is based on Scout, Delta, Thor-Agena, Atlas-Agena, Centaur, and Saturn. Payload capabilities for these various vehicles are summarized in Table 1.

The Scout can launch 150-pound satellites into low altitude earth orbits, or small probes to a distance of several earth radii. The Atlas-Agena-B can launch one or two tons into a low earth orbit, half a ton into a highly eccentric earth orbit, or moderate size payloads to lunar or planetary distances. Starting with the Centaur, the NASA vehicles will rely heavily on the use of liquid oxygen and liquid hydrogen technologies. The Centaur will be able to launch heavy payloads into high altitude earth orbits, for example 24-hour orbits at 35,900 kilometers; or land moderate payloads on the moon. The first version of the Saturn, called Saturn C-1, will permit launchings of 20,000-pound payloads into lower altitude earth orbits.

A later version of Saturn, called Saturn C-3, would quadruple the C-1 capability in an earth orbit and could land five tons on the lunar surface. Even the C-3, however, is not adequate for the manned lunar landing unless a number of them meet in an earth orbit and form a larger spaceship. The Nova is a vehicle conceived for a payload capacity large enough to make a manned lunar landing by direct ascent from the earth and to return.

Although less spectacular than the giant space vehicle, the sounding rocket is important, especially for investigations within the earth's atmosphere. For rocket sounding, NASA has a long

NASA and Space

list of rocket vehicles ranging in altitude capabilities from less than 100 kilometers to many thousands of kilometers.

Supporting the development of necessary vehicles is a strong program of research and development of chemical engines. Oxygen-hydrogen technology is receiving special attention. Jointly with the Atomic Energy Commission, NASA is supporting the development of nuclear propulsion.

SUPPORTING RESEARCH AND TECHNOLOGY

The NASA program includes many activities in support of its aeronautics and space programs, including work on electrical propulsion, such as ion and plasma engines. Research and development on power supplies range from small capacity supplies, needed for smaller satellites and spacecraft, to high capacity supplies needed for deep probes and electrical propulsion devices. Nuclear, chemical, and solar power supplies are being developed. The nuclear development is being done by the AEC. Both direct and indirect conversion methods for solar power supplies are being worked on. New and improved techniques for guidance are being investigated. Bearings suitable for use in space are under development.

The program of basic research embraces the physical, chemical, and aeronautical sciences; materials, mathematics, and astronautics. It ranges from laboratory investigations to experiments in deep space at great distances from the earth. Some of it is done within the organization. Much of it is done by universities and other research organizations.

The Mercury program is a first step toward the long-range objective of making possible direct exploration of other planets. The Mercury effort is designed to determine how much a man can do while orbiting the earth in a satellite. In the Mercury flights the man is not simply an inert passenger, but has to accomplish during the flight many different tasks that will test thoroughly his ability to perform effectively in orbital flight. The Apollo program, already underway, is the next step toward the over-all objective of manned exploration of space. Apollo is intended to achieve the national goal stated by President Ken-

TABLE 1
Launch Vehicle Summary

Vehicle	Stage	Propellant	Stage Weight (lbs.)	Thrust (lbs.)	Maximum Diameter (ft.)	Height, Less Payload (ft.)	Mission Payload (pounds) 550 km. Orbit	Mission Payload (pounds) Escape	First NASA Launch
Vanguard	1	LOX/Ker	17,600	28,000(S.L.)	3.75	72	21.5 (elliptic orbit)	......	March 17, 1958 (Navy)
	2	UDMH/WIFNA	4,200	7,700					
	3	Solid	430	2,200					
Atlas Able	1	LOX/RP	260,000	367,000(S.L.)	10	104	......	360	No successful launches
	2	LOX/RP		80,000					
	3	WIFNA/UDMH	4,600	7,700					
	4	Solid	525	2,800					
Jupiter C	1	LOX/Hydyne	65,000	83,000	5.8	68.5	30	......	January 31, 1958 (Army)
	2	Solid							
	3	Solid							
	4	Solid							
Juno II	1	LOX/RP	120,000	150,000(S.L.)	8.8	77	95 (max. load)	15	Dec. 6, 1958 (Army)
	2	Solid	750	15,000					
	3	Solid	200	4,000					
	4	Solid	60	1,600					
Scout	1	Solid	23,600	103,000(S.L.)	3.3	65	150	......	July 1, 1960
	2	Solid	9,600	62,000					
	3	Solid	2,700	13,600					
	4	Solid	525	2,800					

Vehicle	Stage	Propellants	Thrust					Date
Delta	1 2 3	LOX/RP WIFNA/UDMH Solid	150,000(S.L.) 7,700 2,800	107,000 4,600 525	8	92	500	60	May 13, 1960
Thor Able	Nominally same configuration and performance as Delta except for a less refined guidance system								March 3, 1959
Atlas	1 2	LOX/RP LOX/RP	360,000(S.L.)	260,000	10	97	......	......	July 29, 1960
Redstone	1	LOX/RP	78,000(S.L.)	66,000	5.9	83	Ballistic	......	Dec. 19, 1960
Thor-Agena-B	1 2	LOX/RP IRFNA/UDMH	165,000(S.L.) 15,000	107,000 14,000	8	86	1,600	......	CY 1962
Atlas-Agena-B	1 2 3	LOX/RP LOX/RP IRFNA/UDMH	367,000(S.L.) 80,000 15,000	260,000 14,000	10	98	5,000	750	CY 1961
Atlas Centaur	1 2 3	LOX/RP LOX/RP LOX/H$_2$	367,000(S.L.) 80,000 30,000	260,000 30,000	10	105	8,500	2,300	CY 1961
Saturn(C-1)	1 2	LOX/RP LOX/H$_2$	1,500,000(S.L.) 90,000	....	21.6	125	20,000	......	CY 1961 CY 1963
Saturn(C-3)	1 2 3	LOX/RP LOX/H$_2$ LOX/H$_2$	1,500,000(S.L.) 800,000 90,000	....	26.6	......	80,000	30,000	......

nedy on May 1, 1961: "I believe that this nation should commit itself to achieving the goal, before this decade is out, of landing a man on the moon and returning him safely to earth."

A vigorous effort to develop practical applications for space knowledge and technology is under way. The meteorology program includes, at the present time, the Tiros and Nimbus satellites. Tiros is the first step toward an operational weather satellite. Nimbus is a second step, with the satellite stabilized with respect to earth. Tiros and Nimbus satellites are primarily research vehicles, laying the groundwork for taking cloud pictures and measuring atmospheric radiations, both important to an understanding of the behavior of the total atmosphere. From Tiros and Nimbus will come the future operational systems.

SPACE SCIENCE PROGRAM

Sounding rockets, satellites and space probes provide unique opportunities to attack problems of physics, geophysics and astronomy. The space sciences program is intended to emphasize the interdisciplinary aspects of the problems. The objectives include: investigation of the earth and its atmosphere, and the influence of the sun on the earth; study of the nature and history of the solar system; the search for life in the solar system; extension of astronomical researches to new wave lengths; and contribution to the study of cosmology. In space science, sounding rockets, satellites, and space probes, plus all the conventional techniques of laboratory research, are devoted to an exploration of the earth's atmosphere and of outer space.

The vehicles available at the start of the NASA program could handle the launching of only relatively small satellites. As a consequence, the number of experiments that could be accommodated was also small. Because of this and the long leadtimes involved in preparing a virtually tailor-made satellite, the berths for experiments were quickly signed up several years in advance. The NASA schedule for space sciences experiments is shown in Table 2.

Sounding rockets provide a means for direct measurements in the earth's upper atmosphere and at its edge. With sounding

TABLE 2
NASA Space Sciences Program

Schedule of Major Launchings

	1961	1962	1963 on
Sounding Rockets	85	100	100 per year
Small Satellites	8	6	Several per year
Geophysical Observatory (eccentric orbit)			1 per year
Geophysical Observatory (near earth polar orbit)			1 per year
Solar Observatory		2	1 per year
Planetary and Interplanetary		2	2 each period to Mars and to Venus
Lunar and Interplanetary		3	Several per year
Interplanetary			Occasional special missions

	1963	1964	1965	1966 on
Astronomical Observatory	1	1	1	1 per year

Experiments for launchings to the left of the heavy line are already fixed and under preparation; those to the right represent open opportunities for new and advanced experiments. Sometime after 1965, recoverable, high-altitude sounding rockets will be flown, and an interplanetary vehicle may be sent toward Jupiter. Early lunar flights will be with Atlas-Agena boosters; later flights will use Saturn and Centaur (hydrogen) engine configurations. Early planetary flights may follow this pattern, using Saturns at first, Centaurs later. The eccentric geophysical and astronomical observatories will be launched with Atlas-Agenas, later with Centaurs. The polar orbiting geophysical observatories will be launched with Thor-Agenas. The first two solar observatories will be launched with Thor-Deltas; later models will be launched by Thor-Agena-B's. Later interplanetary work will include near sun probes, and space probes which travel out of the plane of the ecliptic.

rocket experiments, one can investigate the earth's atmosphere and ionosphere, conduct meteorological soundings above balloon altitudes, make observations of the sun and stars at wave lengths that do not reach the ground, and carry out cosmic ray experiments. Sounding rockets may be equipped with recovery packages to retrieve emulsions or other specimens exposed to high altitude conditions, or they may use radio telemetering for obtaining the desired data. Sounding rockets may be used to make a preliminary run of an experiment that is later to go into a satellite or a deep space probe. It is NASA policy to encourage, and sometimes to insist on, such preliminary sounding rocket experiments before the satellite or space probe phase of a project. In general, about one year must be allowed for the preparation of a sounding rocket experiment. Occasionally, if the experimental equipment is particularly complicated, the required time may be longer. For the conduct of sounding rocket experiments accepted into the NASA program, NASA will provide assistance and funds to get flight hardware built by experienced contractors.

RESEARCH SATELLITES

The small satellites in the present NASA program will be launched by the Scout vehicle. These satellites will be used both for the international satellite program and for experiments having special orbital or other requirements. It takes about two years to prepare a Scout type satellite, following the determination of what the payload is to carry. In general, Scout satellites will be prepared under the guidance and direction of the Goddard Space Flight Center. Here again, the actual preparation and construction may be done by a contractor.

When a satellite payload has been determined and assigned to the center, the center names a project manager and a project scientist. The project manager forms a working group consisting of the project scientist, the experimenters, contractors' representatives, and experts on tracking, telemetering, satellite construction and testing. The duty of the project manager, who is in overall charge of the preparation of the satellite, is to see that the satellite is properly prepared and tested, suitably mated to the launching vehicle, that it meets all necessary deadlines, and that the engineering satisfies the legitimate requirements of the scientific experiment. In this last matter, the project manager is assisted by the project scientist, who acts as representative of the experimenters. The experimenters are encouraged to spend as much time as they can at the center so that the requirements of their experiments are made clearly known to the satellite engineers. The working group is so organized that the experimenters have direct access to the project manager to make their needs known directly to him.

Each satellite requires an extensive operation to put it into orbit. The preparation, launching, tracking and related operations all add up to a tremendous effort involving many hundreds or thousands of people. Each launching requires the full use of a launching pad and associated complex. Therefore it becomes prohibitive to attempt to accomplish the space research program entirely with small satellites. To do so multiplies in an unreasonable way the number of necessary large-scale operations per year. As a consequence, it is necessary to carry out as much as possible of the space research program in larger, more standardized satellites, reserving the small tailor-made satellites for those experiments that have special requirements.

NASA is developing several types of observatory satellites to handle the bulk of its satellite research. These are geophysical observatories, solar observatories, and astronomical observatories. They are stabilized satellites, with standardized power supplies and standardized telemetering. Command links will be provided for sharing among the experiments the telemetering capacity of the observatory, and for the performance of other necessary functions. The long-range plan is to fly one of each type of observatory per year. As much flexibility as possible is to be built into each satellite so that it is not necessary to re-engineer the basic observatory for each new flight.

Individual experiments often have special requirements. With imaginative engineering of these observatories sufficient flexibility can be achieved, along with design standardization, to meet the requirements of the various experiments. It is intended that the leadtime for the preparation of experiments to go aboard the observatories will be a year (or less for the simpler instrumentations). Thus, it would be possible for an experimenter to follow up an exciting discovery with improved instrumentation for the succeeding observatory. Of course, when the experimental equipment itself gets exceedingly complicated, as may well be the case with the astronomical observatories, the leadtime of one year may be greatly increased.

The orbiting geophysical observatories will be launched into orbits designed to investigate the earth's atmosphere and the earth's environs in space. One type of trajectory will be a near earth polar orbit; another an eccentric orbit reaching 100,000 kilometers and beyond, depending upon the capability of the launching vehicle. Both the solar observatory and the astronomical observatory will be launched into near earth orbits of low inclination, above the earth's atmosphere, but below the Van Allen radiation belt.

The pointing accuracy of the astronomical observatory will be the most precise; that of the geophysical observatory least precise. The pointing accuracy in the case of the astronomical observatory will be seconds or fractions of seconds of arc; that of the solar observatory, minutes of arc; that of the geophysical observatory, a few degrees. Moreover, the orientation of the astronomical observatory will be changeable on command, so that the telescopes can be pointed at several regions of the sky.

The first orbiting solar observatory is scheduled for launching in 1962. The satellite will weigh, when completely equipped with instruments and power supply, about 350 pounds. Solar cells and internal batteries will be used for the power supply. The satellite consists of a large flywheel with extended rotating arms to stabilize the satellite in space. A fixed arm contains instrumentation that points steadily at the sun throughout the course of the orbit. Solar pointing control elements are connected with a compressed gas jet in such a way as to precess the gyroscopic element into proper alignment. Experiments which must view the sun continuously are placed in the fixed arm. Experiments which may be permitted to scan around the celestial sphere are placed in the outer rim of the wheel. There is opportunity for experiments to go on subsequent solar observatories. Proposals and suggestions will be welcomed by NASA.

The first geophysical observatory will be launched in 1963 or 1964. It will be placed in an eccentric orbit extending to roughly 100,000 kilometers from the earth. These satellites will provide a continuing opportunity to investigate the earth and its atmosphere and cislunar space.

The first astronomical observatories will be launched in 1964, 1965, and 1966. The first of these will carry sky survey experiments prepared by the Smithsonian Astrophysical Observatory and the University of Wisconsin. The second will carry an ultraviolet stellar spectroscopy experiment prepared by the Goddard Space Flight Center. The third will carry a precision spectroscopy experiment prepared by Princeton University. Experiments for the later astronomical observatories are yet to be determined.

MOON AND PLANETS

The NASA program includes investigation of the moon. The first flights will be exploratory in nature with the prime objective of taking good resolution pictures. Measurements of the radioactivity of the lunar surface to determine the composition of the lunar crust will be made. By means of equipment landed on the surface, the character of the surface and the seismic activity of the moon will be measured. As capabilities of vehicles improve, precision orbiters about the moon and soft-landed packages on the moon will continue the investigation in greater detail.

The precision orbiting flights will be used to map out the visual appearance of the moon and to continue the radioactivity measurements. Those which land will be used to study physical and chemical properties of the surface of the moon; to measure the thermal regime of the moon, including the heat flow from the interior; to continue long term observations of solar particles, the solar "wind," and the interplanetary magnetic fields. In time, samples will be collected for return to Earth, and roving vehicles will explore the lunar surface. The payloads up through 1962 are being prepared. Payloads to follow afford opportunities for further, more advanced experimenting, particularly in the area of surface exploration.

Early flights to the vicinity of Mars and Venus will be fly-bys. That is, the payload will pass by the planet at a probable distance of several planetary radii. These flights will be instrumented for both interplanetary and planetary measurements. For Venus flights, a prime objective will be to determine the surface temperature of the planet and to learn as much as possible about the composition and nature of the planet's atmosphere. Photography of the surface of Mars will be of primary importance, as well as a search for organic substances.

As soon as possible, entry into the planetary atmospheres will be attempted to study atmospheres in detail, to investigate planetary ionospheres, look for the existence of radiation belts, and to lay the groundwork for a closer examination of the planet when landings become possible. As soon as it does become possible to place equipment on the surface, planetary observatories will be landed to continue study of the lower atmosphere, and to investigate the surface in detail. On Mars, especially, there will be a search for primitive forms of life. The experiments for the 1962 flights which will be directed toward Venus have been chosen, but those for the flights to Mars and Venus in 1964 are not yet entirely decided. There are exciting opportunities for conducting experiments on planetary and interplanetary vehicles in 1964 and later.

Planetary and interplanetary vehicles bring to light a host of difficult engineering problems. There is the matter of building in enough power supply to maintain communications between the spacecraft and the earth. One must also provide a suitable temperature control in a vehicle whose distance from the sun is continuously changing. Suitable antennae must be provided for observing the planetary object in question, for receiving commands and messages form the earth and directing answers back. The complications that arise make the development of these spacecraft extremely challenging; very long leadtimes are required. For the Saturn vehicles which will be produced in the latter half of this decade the engineering leadtimes will be quite long. However, it is the NASA intention that the spacecraft engineering be sufficiently flexible that the leadtimes for scientific experiments can be kept to a minimum. It is recognized that good experiments would be difficult to achieve if many years had to elapse between the conception of an experiment and its execution. It is hoped that leadtimes for scientific experiments can be kept down to one or two years.

The lunar and planetary flights will also attempt interplanetary measurements. In addition, there will occasionally be special interplanetary missions such as solar probes to within the orbit of Mercury, or a probe to be flown outside of the earth's ecliptic. Such missions offer exciting possibilities for space research.

Experimenters are expected to publish data collected in NASA programs in analyzed form in the open literature. At the same time, NASA publishes results from its projects. On occasion, these publications may be reprints of the articles in the open literature. NASA publication affords the opportunity for a more extensive and more detailed account of the project. Moreover, NASA can undertake to publish tables of reduced data that would not be accepted by, and could not be handled by, the professional journals.

It is NASA policy that the experimenter should have sole use of the data obtained from his experiment for a period of time long enough to accord him his rights as the conceiver of the experiment. On the other hand, because a tremendous amount of the nation's money and the supporting work of a large team of people go into the experiment, and because the data may be of widespread interest to the scientific community, NASA arranges with the experimenter for general release of the data after an agreed-on interval.

With the agreement of the experimenter, the scientific community is sometimes invited to record the signals from a given satellite, and the telemetering code and calibrations are made available to interested scientists so that they can collect, reduce, and analyze the data as though they were themselves performing the experiment. The disposition of experimental data is worked out in prior arrangements with the experimenting scientists.

PARTICIPATION IN THE NASA PROGRAM

NASA extends an invitation to the scientific community to make use of the research opportunities afforded by the scheduled satellites and space probes (see Table 2). This invitation has also been extended to scientists abroad. Selection of experiments for flight in NASA vehicles depends on the pertinence of the proposal to the over-all NASA program, on its competence and its merits relative to similar competing proposals, and on the actual requirement for a sounding rocket or space vehicle for the performance of the proposed experiment. The practicalities of budget, scheduling, and the amount of NASA laboratory and field support required are also considerations.

Accepted proposals will be provided with space aboard appropriate satellites, space probes, or sounding rockets. Contracts or grants include necessary funds for subcontracting the engineering of experimental hardware. To insure success and to protect the interests of the experimenters, NASA prescribes pre-flight testing for prototype and flight equipment. NASA conducts the flight acceptance tests and assists in the conduct of design development tests. The integration of the experimental instrumentation into flight payloads will be accomplished either by NASA centers or by systems contractors under the supervision and direction of NASA.

DEPARTMENT OF DEFENSE SPACE PROGRAM

A. G. WAGGONER

In his appearance before the Senate Committee on Aeronautical and Space Sciences on June 1, 1961, the Deputy Secretary of Defense discussed the relationship between the National Aeronautics and Space Administration and the Department of Defense as follows: "It is important to recognize and understand that there are some space applications which are distinctly military; others which are of mutual interest for civilian as well as military use. Still other missions are, at present, primarily of civilian or scientific interest. However, these latter will provide fundamental knowledge which may provide a basis for military application at some future date. Similarly, out of military programs, technical data is produced which has civilian application." These few words defined the role of the Defense Department in the national space effort. In this era of massive technological advance, the military posture of our country is critically dependent upon the awareness and effective use of national resources.

Development of weapons systems, such as ballistic missiles, to meet military objectives provided much of the early technology required to initiate the exploration and exploitation of space. Thus, when it became apparent to scientists in 1954 that a satellite would assist in achieving the objectives of the IGY, it was agreed that a satellite launching vehicle, the Vanguard would be developed by the Navy to meet the needs defined by the United States National Committee for the IGY, a committee of the National Academy of Sciences.

Some studies of hypothetical satellite systems had been made prior to this time, but the Vanguard constituted the first serious effort for the production of hardware. When that program encountered technical difficulties, still another vehicle, the Army

Redstone re-entry vehicle, which had been developed to provide information on the nose cone test of the ballistic missile, was used to place the first United States satellite into orbit. From this meager start, the Defense Department programs have contributed knowledge on propulsion systems, guidance techniques, design characteristics and vehicle hardware to move our expanding space effort along as fast as practicable.

With the successful launching in February, 1958, of the first United States satellite, Explorer I, a vast panorama of opportunities unfolded, opportunities which promised to affect both the military and civilian technology. Early satellite studies, some dating back to the late 1940's, were dusted off, brought up to date, and proposed for development and use. Scientific experiments and studies on navigational satellites, reconnaissance and early warning satellites, communications satellites, meteorological satellites, and many others were initiated, all aimed at exploring or understanding the space environment or enhancing our present capabilities. In 1958, the Secretary of Defense established the Advanced Research Projects Agency (ARPA) within the Department of Defense and assigned as the initial task of that agency the formidable job of reviewing the many programs and proposals to establish an over-all plan for the Defense Department space effort and to initiate the appropriate programs. All of the space programs proposed and those then under way were transferred from the services and placed under ARPA program control and funds.

While practically all of the present Defense Department space programs may be traced back to the ARPA initial programming and direction, it should be noted that the basic test for such a program was established by Congress in the formulation of the National Aeronautics and Space Act of 1958 in which the space mandate was given to the NASA, except for activities primarily associated with development of weapons systems, military operations, or defense of the United States, which are the responsibility of the Department of Defense.

Implementation of the act caused a major reorientation of the Defense Department's space effort. Organizations and programs primarily oriented in the direction of the NASA research, space

Department of Defense Space Program

exploration, and exploitation objectives were transferred to NASA. Other programs of specific interest to the Department of Defense were screened carefully to assure compliance with the intent of the act. In areas of joint interest, specific agreements defining the responsibilities of each agency have been formulated, with continuous monitoring by the Department of Defense–NASA Aeronautics and Astronautics Co-ordinating Board and its subdivisions. Further technological developments in the areas of guidance and propulsion, continued use of military launching vehicles for Defense Department research, and the common support by the national launching ranges of both the Defense Department and NASA programs have all emphasized the need for close co-operation between these agencies for the utilization of available resources and exchange of technical advancements.

At present, the Department of Defense programs can be grouped into three broad categories: first, research to achieve more fundamental knowledge and techniques necessary to assure the desired reliable performance and survivability of military satellites in space; second, developmental projects aimed directly at utilization of space for military missions; and, third, support of other parts of the national space program.

DEPARTMENT OF DEFENSE SPACE RESEARCH

Probably the best known space research and techniques project is the Air Force Discoverer. This project, initiated by ARPA, had its first flight in February, 1959. As of November, 1961, thirty-five Discoverer satellites had been launched, twenty-four of which achieved orbit. A number of technological firsts may be credited to this program. Some examples are space capsule stabilization, satellite engine restart after coast period, achievement of precision circular orbit, and initial demonstration of a spaceborne recovery technique with the recovery of the data capsule from Discoverer XIII on August 10, 1960.

The Discoverer satellites have also been particularly useful in determining the characteristics of the space environment. Typical of the experiments performed are the use of samples of metals, such as silicon, bismuth, iron, lead, magnesium, nickel, and titanium aboard Discoverer XXXV as "chemical detectives" because of their individual sensitivities to impacts from various types of energetic particles. Each type of metal displays a distinctive change upon impact from a certain type of radiation. Changes in the materials indicate the type of radiation responsible for the change. For example, the change of the bismuth sample into a particular isotope of bismuth would indicate the impact of neutrons.

The Discoverer series uses as a first-stage rocket the standard Thor intermediate range ballistic missile modified to accommodate the Discoverer second stage. This second stage, the Agena, separates from the first stage at the end of the boost period, coasts for 15 seconds and then is propelled into a polar orbit by the second stage rocket engine. The total weight in orbit is about 1,700 pounds of which about 300 pounds consists of instrumentation, telemetry, re-entry capsule, retrorocket, and recovery aids. The height of the combined stages at time of the launching is 78 feet and the weight is over 100,000 pounds.

Complementing the research effort carried on by the Discoverer program are the many "piggy back" research satellites launched as passengers on the vehicles carrying other primary payloads. The Army, the Navy, and the Air Force plan to make extensive use of this technique since it permits collection of the desired data at a minimum of cost to the scientific organizations interested in these projects. Typical are the Navy Greb solar radiation measuring satellites launched with the Navy Transit navigational satellites. All such fundamental research efforts are co-ordinated with NASA.

Indirectly supporting the space effort are the Army-Navy-Air Force upper atmospheric rocket probe programs and the manned research aircraft programs such as the X-15 and Dynasoar. The X-15, being developed by the Air Force and NASA, is a rocket-powered maneuverable aircraft. Design objectives are a speed of 4,000 miles per hour and an altitude of fifty miles. At this time, the speed objective has been accomplished and achievement of the altitude objective is expected in the near future.

The Dynasoar is being developed by the Air Force. In contrast to the X-15, the Dynasoar is a rocket-boosted, delta-winged glider which will fly from speeds slower than the landing speed of jet aircraft to more than twenty-five times the speed of sound. Through the Dynasoar, it is hoped to determine the flight parameters for controllable vehicles operating from satellite recovery speeds to the speeds now being explored by the X-15. The initial tests will achieve suborbital velocities.

The boost-glide heat resistant vehicle will be tested at Cape Canaveral, Florida. The re-entry vehicle is designed so that its pilot retains full control of his maneuvers once he has re-entered the atmosphere at a speed of about 15,000 miles per hour. He will be able to make visual observations, operate instruments, and communicate with associates on the ground. He also will be able to select different landing sites because of Dynasoar's ability to change from ballistic to glide flight once it has entered the atmosphere. In later orbital tests, the re-entry vehicle would perform a skip-glide re-entry path, finally landing at a conventional airbase. Completion of the current Dynasoar program promises to provide the desired experimental data for a manned, controllable vehicle which can operate at speeds ranging from those attained by satellites (10,000 miles an hour) to those required for landing at a conventional military airport (several hundred miles an hour).

The upper atmospheric research program, carried on in close co-operation with other government agencies, such as NASA and the Weather Bureau, has the objective of achieving a better understanding of the near space environment. Some examples are communications studies of effects of solar disturbances on the ionosphere, meteorological studies of interactions of the lower and upper atmosphere, and radiation levels associated with varying solar activities. In addition, related bioastronautical investigations on weightlessness and other space environmental properties assist in the assessment of the proposed military uses of space.

MILITARY SPACE PROJECTS

Development projects having specific military missions, can be subdivided into three areas: (1) Information-gathering satellites such as the military warning and surveillance types (Samos, Midas), (2) inspection satellites (Saint rendezvous and Vela Hotel nuclear detection satellites), (3) support satellites (Transit navigation and Advent communications satellites and the West Ford communications experiment). Information and surveillance satellites are being developed to permit the earliest possible detection of a missile attack on the United States. When Midas is developed and in operation, infrared sensors located in several of these satellites should be able to detect the heat from exhaust flames during a launch occurring anywhere in the world. Such information telemetered instantaneously to data read-out stations will permit an earlier warning to our retaliatory forces, thereby increasing considerably the probability of their survival.

The Midas satellite is boosted into a polar orbit by a vehicle consisting of a modified Atlas intercontinental ballistic missile with an Agena B vehicle as a second stage. In the launch sequence the Agena vehicle separates at the end of the Atlas boost period and coasts briefly before the initial rocket engine ignites. After a brief burning period sufficient to place the satellite in an elliptical orbit the Agena again coasts to the apogee of the orbit, a point in space about halfway around the earth from point of launch. At this time the Agena engine is restarted, placing the Midas in a near-circular high altitude orbit. The height of the Atlas-Agena combination at time of launch is about 98 feet, with a launch weight of approximately 270,000 pounds. The final orbital weight is about 3,500 pounds.

The inspection satellites pose a number of interesting engineering problems. The objective of the Saint program is the development of an unmanned maneuverable inspection satellite. Since the ability to inspect unidentified satellites is desired, orbital control and various inspection systems such as radar and television, and satellite-to-ground communication systems are required. Especially in the Saint development it is necessary to solve aspects of the rendezvous problem. NASA also is developing specific techniques for solving the problem of rendezvous with co-operative satellites for possible refueling and transfer operations. While the objective of inspecting an unco-operative satellite is different, it is obvious that there are many common technical elements, and the two programs are closely co-ordinated. Perhaps the more fundamental problem is associated with the final selection of the inspection sensors. It is apparent that the sensors must be non-destructive even though a non-co-operative satellite is involved. It is expected that the Atlas-Agena vehicle will be used as the launch vehicle for this program.

Vela Hotel presents a problem of longer-range detection. Its objective is the achievement of a satellite system to detect nuclear explosions in space. Again the Atlas-Agena vehicle will provide the boost capability. A principal objective of the Vela Hotel research program is to provide information leading to a working system capable of supporting international agreements which may be developed by the Geneva nuclear test control discussions.

In the study of support satellites, there have been successful advances achieved by the Navy in development of the Transit navigational satellite. Achievement of a prototype high capacity computer-type memory and operational radio frequencies has provided a minimum operational capability. Because of the long-term power requirements for this satellite a SNAP (Special Nuclear Auxiliary Power Systems) nuclear generator power source is being tested with this satellite. When fully operational, the Transit navigational satellite will furnish necessary information for ships to determine their positions. Ships will require radio receivers and computers to use the planned world-wide all-weather system.

Another major development in support satellites is the Advent communications satellite, under development by the Army. This development is an attempt to supplement the present Department of Defense global communications net required for the active deployment and control of our armed forces. The present plan calls for placing the satellite in a 24-hour synchronous orbit using the Atlas-Centaur launch vehicle. The program is closely co-ordinated with NASA, to avoid unnecessary duplication between the Department of Defense program and the NASA commercial communications program, and to assure payload and launch vehicle integration. Project West Ford, an experiment in the support category, is being performed to determine the practicality of establishing a belt of orbiting needles—dipole magnets—for communication purposes.

DEPARTMENT OF DEFENSE SUPPORT OF NATIONAL SPACE PROGRAM

In considering the third basic category of Department of Defense support of other parts of the national space program, one must recognize the basic broad self-sustaining characteristics of the armed services. The Department of Defense is participating with NASA in a national launch vehicle program. Under the terms of this program, the responsibility for development or procurement of each launch vehicle has been identified and assigned. In general, Department of Defense work relates to components developed under one of the major missile programs and adapted for use in the space effort. Thor and Atlas boosters fall into this category. Also within this same framework is the decision for the Defense Department to develop an all solid propellant satellite launch vehicle for use by NASA or the Defense Department.

Perhaps the largest well-defined support areas are the national ranges and associated launch pads; specifically, the Atlantic Missile Range operated by the Air Force, and White Sands Missile Range, operated by the Army, and the Pacific Missile Range, operated by the Navy. These three ranges, representing an investment of over 1.5 billion dollars, have provided ready support in the form of flight and test procedures and world-wide bases for location of the various satellite tracking stations. It has been agreed that the range support activities are available to any United States program requiring their capabilities.

In some tests requiring certain capabilities, special arrangements are made. In the Project Mercury tests, naval aircraft units are available to assist in the performance of the test operations.

FUTURE TRENDS

There is little doubt in the minds of those closely connected with the Defense Department space program that there is a continuing and growing effort in this area. As more data become available to assess the opportunities within the space environment, new requirements will be defined and additional systems developed. The introduction of new techniques and a better understanding of the environment should provide greatly sophisticated inspection capabilities both on earth and beyond the sun. Controllable interceptors utilizing new power plants in the space environment will permit control of space by agreement—or by force.

In summary, the Defense Department's space program has evolved from its inception in the late 1950's through several organizational and policy steps into a program consisting of projects to support and enhance the over-all capabilities of the services. These projects are aimed at meeting specific military objectives, and at supporting research to define further military requirements.

SPACE PROGRAMS OF OTHER NATIONS

JOEL ORLEN

Some indication of the extent of growth in space research can be found in the participation of national scientific institutions in the activities of the Committee on Space Research (COSPAR) established by the International Council of Scientific Unions (ICSU). COSPAR's Second International Space Science Symposium held at Florence, Italy, in April 1961 was attended by nearly 300 scientists from 27 countries, and 117 scientific papers were presented by scientists from 14 countries. Membership in COSPAR has risen to eighteen representatives of national scientific groups from the original seven that were active in the International Geophysical year (IGY) rocket and satellite program. The eighteen scientific institutions currently represented are Argentina, Australia, Belgium, Canada, Czechoslovakia, France, the German Federal Republic, India, Italy, Japan, the Netherlands, Norway, Poland, South Africa, Switzerland, the United Kingdom, the Soviet Union, and the United States. In addition, Israel, Pakistan, Peru, and Sweden, countries whose national scientific institutions are not now officially represented in COSPAR, have initiated space research programs. While only the United States and the Soviet Union have launched satellites, other countries are designing experiments to be carried aboard satellites, as well as launching rockets, tracking satellites, receiving satellite telemetry, conducting ground-based experiments relevant to those carried out in space. These countries are also contributing to theoretical analyses of data from rockets and satellites.

The most striking evidence of the world-wide growth of space research can be found in a review of the scientific work in progress in each of these countries, and the following summary is based primarily, but not exclusively, on reports submitted to COSPAR by national scientific institutions from each country, and on a survey of the scientific papers presented at COSPAR's two international space science symposia.

THE SOVIET UNION

The Soviet Union has a series of notable achievements in space exploration: first artificial earth satellite (Sputnik I, October 4, 1957), artificial satellite in a lunar orbit (Lunik I, January 2, 1959), lunar impact (Lunik II, September 12, 1959), photographs of the far side of the moon (Lunik III, October 4, 1959), and the first manned space flight (Vostok I, April 4, 1961). The following summary of the Soviet scientific space effort is based primarily on Soviet reports to the world data centers and to COSPAR. In addition, scientific papers presented by Soviet experts to COSPAR symposia and review articles of Soviet scientific journals have been used.

From Sputnik I to the end of 1961, the Soviet Union has launched thirteen earth satellites and three space probes. The majority of the satellites have been devoted primarily to developing the capability of placing a man into orbit around the earth. However, a number of them were instrumented for scientific measurements and observations. These included the three satellites launched during the IGY, Sputniks I, II, and III. They carried instruments for measuring density, temperature, and structure of the upper atmosphere, the composition of the ionosphere, frequency of micrometeorite impacts, corpuscular radiation, and the lower boundaries of the belts of trapped radiation. Studies of changes in the parameters of the orbits of these satellites showed that the increased density of the atmosphere at medium to high altitude is closely correlated with solar activity. Since the IGY itself coincided with the maximum phase of several solar activity cycles (5, 11, and 100 years) the atmosphere at satellite altitudes was at its maximum density and temperature during all or part of the lifetime of the earliest satellites. It turned out that the lifetime of Sputnik III (May 15, 1958, to April 6, 1960) was significantly greater than had been predicted on the basis of maximum atmosphere density, apparently because, after the maximum phase, the density had already begun to diminish.

In 1959 the Soviet Union launched Lunik III, which circled the moon and went into an orbit around the earth-moon system. This was the satellite that transmitted back to the earth pictures of the reverse side of the moon. The Soviet Academy of Sciences has published these results in *An Atlas of the Reverse Side of the Moon,* which contains thirty large-scale reproductions of detailed photographs and a catalogue of the newly identified features of the hitherto unseen portion of the moon. The three satellites the Soviet Union launched in 1960, Sputniks IV, V, and VI, were largely devoted to developing the capability of launching a man into space. They did, however, carry scientific instruments for measurements similar to those made by the first three Sputniks. A global map of cosmic ray intensity distribution was recorded with areas of maximum intensity in the region of the Taimyr peninsula, in North America, and in the South Atlantic. The six satellites launched in 1961, Sputniks VII–IX and Vostoks I and II, were once again devoted largely to the man-in-space program. Other scientific results from these satellites, if any, have not yet been reported.

The Soviet effort to launch a man in space included experiments aboard nine of the thirteen earth satellites. It began with Sputnik II, launched November 3, 1957, and includes Sputniks IV, V, VI, VII, IX, and X and Vostoks I and II. Sputnik II carried the dog, Laika, and medical and biological experiments, the results of which demonstrated that animals can function under conditions of space flight.

Sputniks IV, V, and VI, launched in 1960, tested life-support systems (air-conditioning, regeneration, and thermal regulation) and re-entry techniques. Sputnik IV carried a hermetically sealed cabin with a payload to simulate a man's weight. Sputnik V consisted of a space cabin containing two dogs and other biological specimens along with life-support equipment and an instrument section. The cabin was catapulted from the remainder of the satellite and recovered. Control and orientation equipment was tested for several days during flights of all three satellites. Tele-

Space Programs of Other Nations

vision equipment for direct observation of the animals also was developed for these flights. No significant anomalies of basic physiological functions were recorded, but transient functional shifts in biochemical indices of blood were observed, as well as some changes in immune properties. These were assessed as a general non-specific stress reaction to the influence of a combination of factors during flight. Investigation of the tonus of peripheral vessels did not reveal any changes. Examination of the dogs by veterinarians from a general physiological and clinical point of view also revealed no significant changes. No deviations were recorded when rats carried aboard the satellites were examined for changes in the health and dynamics of conditioned reflex reactions and for changes in the functional status of the central nervous system.

Sputniks VII, IX, and X, all launched in 1961, further developed the spacecraft and life-support systems for the successful manned flights later that year. On April 12, 1961, Vostok I was launched carrying Soviet cosmonaut Yuri Gagarin on the first manned space flight. On August 16, 1961, Vostok II carrying Soviet cosmonaut Gherman Titov, was launched and achieved seventeen revolutions before it was safely returned to earth.

The first Soviet space probe was Lunik I, launched on January 2, 1959. It traveled to within 5,000 to 6,000 kilometers of the moon. Lunik II, which was launched on September 12, 1959, hit the surface of the moon. These Luniks carried scientific instruments for measuring the intensity of primary cosmic rays, X-rays and gamma rays, and for determining the composition of charged particles in space. Scientific results reported include detection of the outer belt of trapped radiation whose maximum intensity occurs at a distance of four radii from the center of the earth; discovery of a system of non-ionospheric electric currents at a distance of about three to nine terrestrial radii; and quantitative measurements of the magnetic field near the earth and in space. No notable magnetic field was recorded near the moon.

A Venus probe, which was launched on February 12, 1961, from Sputnik VIII while in an earth orbit, was placed in a trajectory that was intended to pass near Venus at a distance of about 100,000 kilometers. After February 27, radio contact with the Venus probe could not be established. On March 2, the Soviet press announced that as a result of analyses of trajectory measurements, the Venus probe was continuing along its planned path and that on March 3 it would be 4,152,700 miles from a point on the surface of the earth at latitude 1°15′ south and longitude 65°30′ east. The velocity of the probe at this time was stated to be 4,166 miles per second. It carried instruments for measuring cosmic rays, magnetic fields, charged particles of interplanetary gas, micrometeor impacts and temperature. Prior to the loss of radio contact, data from the instrumentation were being received by the Soviet scientists and are to be published after they are reduced and analyzed.

From the start of the IGY until the end of 1960, the Soviet Union launched 345 rockets. Of these, 318 were meteorological rockets for studies of the upper atmosphere; 112 during the IGY, 46 during 1959, and 160 in 1960. Distribution of these launchings by location were as follows: High latitude observatory on Heiss Island in Franz-Josef Land—113; middle latitudes of the European part of the U.S.S.R.—90; from aboard ship in the Antarctic—43, in the Pacific Ocean—67, and in the Black Sea—5. Nearly all these sounding rockets were equipped with electrical resistance thermometers and thermal and membrane manometers. Extensive observations have been collected on pressure, density, and temperature, and their changes with time and latitude in the stratosphere as well as information on the flow of air streams. Analysis of this material has yielded more precise knowledge on the effects of pressure and temperature in the stratosphere and on the interconnection between the stratosphere and the troposphere. A number of the sounding rockets were also equipped with instruments for optical observations of the luminosity of the sky, and the distribution of ozone.

Twenty-seven of the rockets launched by the Soviet Union were higher altitude rockets (100–470 kilometers), designated as geophysical rockets; thirteen were launched during the IGY, seven during 1959, and seven in 1960. Six of these rockets carried hermetically sealed, spherical, self-oriented capsules and instruments for measuring structural parameters of the atmosphere, for optical observations, and to record physical conditions inside

Space Programs of Other Nations

and outside the capsule. The total combined weight of each capsule with instruments was 367 kilograms.

Seven rockets carried to a height of 210 kilometers two capsules—a detachable nose cone equipped with research instruments and a sealed capsule with experimental animals. The total weight of the scientific apparatus and experimental animals carried by each rocket was 2,200 kilograms. Reliable and safe return was assured by a parachute system. Four rockets launched in 1958 carried scientific instruments and experimental animals to a height of 450–470 kilometers. The total weight of instruments and animals in each of these launchings was 1,500 kilograms. These rockets were stabilized throughout their flights. In a launching that occurred on August 26, 1958, scientific instruments and animals were recovered safely from an altitude of 450 kilometers.

Instruments aboard these rockets measured pressure and density of the upper atmosphere; chemical composition and electron and ion composition of the ionosphere; micrometeorite impacts; and distribution of brightness of the day and night sky. Photographs were taken of the earth's surface and cloud systems. Animals and biological specimens were flown to heights of 470 kilometers to check physiological functions in flight.

The seven geophysical rockets launched during 1960 reached altitudes of 100 kilometers and 200 kilometers. Instruments aboard these rockets provided data on X-ray intensity, solar ultraviolet radiation, scattered solar radiation, night air glow intensity, cosmic and terrestrial gamma radiation, day-night cosmic ray intensity variations, electron concentration, and atmospheric density. Medical and biological observations of dogs and rabbits in flight provided additional data on physiological functions.

A few results from additional geophysical rockets launched in 1961 were reported at the COSPAR meeting which took place in April of that year, particularly in connection with observations made during the period of the solar eclipse of February 15. These rockets were launched in the mean latitude of the European part of the U.S.S.R. to explore the solar corona during the period of total solar eclipse. The rockets carried high altitude capsules, referred to by the Soviets as automatic geophysical stations, to an altitude of 100 kilometers. The capsule was kept in a stabilized position during flights in the lunar shadow. To explore the outer solar corona, photographic and photoelectric methods were used to record both relative and absolute measurements of its glow. Data on the brightness of a definite region of the outer solar corona were obtained but as yet have not been reported.

X-ray and ultraviolet radiations of the solar corona were investigated by measuring the total energy of radiation at wave lengths up to 1,300 angstroms and radiation in the spectral regions 2 to 14 angstroms and 40 to 100 angstroms. Scattered and reflected ultraviolet radiation in the lunar shadow also was investigated. Finally an infrared radiometer was used to study radiation of the earth and atmosphere in the infrared spectral region. Results obtained by all three of these instruments have not yet been reported.

At the COSPAR meeting in April, 1961, the Soviet representative reported briefly and very generally on plans for future work. The program of sounding rockets will be continued to study temperature, pressure, and air streams in the stratosphere. Additional high altitude rockets will be launched, to continue measurements in the upper atmosphere similar to those made during the last two years. Earth satellites will be used for explorations and measurements of cosmic radiation and radiation belts, the geomagnetic field, the ionosphere, radio wave propagation, solar ultra short-wave radiation, the earth's radiation balance, and meteorite particles. Earth satellites will also be used for physiological experiments and for the "realization of space flights," a reference to their man-in-space activity.

ARGENTINA

Argentina is now planning a program of rocket launchings, initially as an extension of its active program of ground-based measurements for the monitoring of cosmic rays and meteorological parameters. Argentine scientists operate a network of four cosmic ray stations, extending from near the magnetic equator to

TABLE 1
Soviet Earth Satellites, Cosmic Rockets, and Spaceship Satellites

Designation		Launch Date and Inclination to Equator	Initial Values			Payload Weight (kg.)
Popular Scientific	Soviet Name		Apogee (km.)	Perigee (km.)	Period (min.)	
Sputnik I 1957 Alpha	First Soviet Earth Satellite	Oct. 4, 1957 64.3°	947	228	96.17	83.6

Systematic radio-technical and optic observation of the orbit. Study of the spread of radio waves in the ionosphere.

Sputnik II 1957 Beta	Second Soviet Earth Satellite	Nov. 3, 1957 65.4°	1,671	225	103.75	508.6

Study of the behavior and the physiological functions of the animal during space flights. Study of cosmic rays. Study of solar radiation in the short-wave ultraviolet and X-ray areas.

Sputnik III 1958 Delta	Third Soviet Earth Satellite	May 15, 1958 65.4°	1,880	225	105.95	1,327

Determination of pressure and the composition of the atmosphere. Measurements of the concentration of positive ions. Measurement of the value of the electric charge of the sputniks and the tensions of the electrostatic fields of the earth. Study of the intensity of solar corpuscular radiation. Measurement of the tension of the terrestrial magnetic field. Study of the composition and variations of the primary cosmic radiations and the distribution of photons and heavy nuclei in cosmic rays. Study of meteor particles.

Lunik I	Cosmic Rocket I	Jan. 2, 1959 1° to ecliptic	Aphelion: 197,200,000	Perihelion: 146,400,000	450 (days)	361.3

Magnetic field of the earth measured and magnetic field of the moon explored. Cosmic radiation studied. Gaseous component of interplanetary substance and of corpuscular solar radiation studied. Meteor particles studied. Formation of artificial sodium comet and observation of it.

Designation		Launch Date and Inclination to Equator	Initial Values			Payload Weight (kg.)
Popular Scientific	Soviet Name		Apogee (km.)	Perigee (km.)	Period (min.)	
Lunik II	Cosmic Rocket II	Sept. 12, 1959 65°	Impacted moon at 5:02:24 p.m. EDT September 13, 1959			390.02

Surface of the moon attained. Cosmic radiation studied. Gaseous component of interplanetary substance studied. Magnetic fields of the earth and the moon measured. Radiation belts around the earth studied. Meteor particles studied.

Lunik III 1959 Theta	Cosmic Rocket III	Oct. 4, 1959 80° just after passing moon	480,000	40,000	15 (days)	435

Flight round the moon of automatic interplanetary station carried out; station carried radio-engineering, photo-TV, and scientific equipment. Reverse side of the moon photographed and picture transmitted to the earth. Continuation of the research made in the previous cosmic rockets.

Sputnik IV 1960 Epsilon	Spaceship satellite I	May 15, 1960 64°9'	369	312	91.8	4,540

Test of life-support systems and re-entry of cabin. Recording cosmic radiation levels. Investigation of cosmic rays.

Sputnik V 1960 Lambda	Spaceship satellite II	Aug. 19, 1960 64°57'	340	306	90.7	4,600

Test of life-support systems and re-entry of cabin. Recording cosmic radiation level. Investigation of cosmic rays.

Sputnik VI 1960 Rho	Spaceship satellite III	Dec. 1, 1960 64°58'	265	187	88.5	4,563

Test of life-support systems and re-entry of cabin. Recording of cosmic radiation. Investigation of cosmic rays. Investigation of solar short-wave radiation. Medical and biological investigations.

Sputnik VII 1961 Beta		Feb. 4, 1961 64°	328	223	89.90	644.1

Radio telemetry system to keep watch on the parameters of the ship's structure and equipment for trajectory measurement. Develop and place heavy weight space vehicles in precise orbit.

Space Programs of Other Nations

Designation		Launch Date and Inclination to Equator	Initial Values			Payload Weight (kg.)
Popular Scientific	Soviet Name		Apogee (km.)	Perigee (km.)	Period (min.)	
Sputnik VIII 1961 Gamma 3		Feb. 12, 1961 65°01'	6,658	6,601	89.7	Not disclosed
Venus Probe 1961 Gamma 1		Feb. 12, 1961 0.3° to ecliptic	Aphelion: 152,000,000	Perihelion: 107,400,000	300 (days)	643.5

Test methods of injecting probe into interplanetary trajectory; test long-range communications; provide measurements of solar system; make physical observations of outer space. Sputnik VIII achieved earth orbit and while orbiting launched an instrumented probe toward Venus. Measure cosmic rays; magnetic fields; charged particles of interplanetary gas and corpuscular sunbeams; record micrometeor impacts; temperature control, telemetry, attitude control, and stabilization systems.

Sputnik IX 1961 Theta		Mar. 9, 1961	249.3	185	Not disclosed	456.2

Further testing of construction of the space ship and systems to insure necessary conditions for manned flight; test influence of cosmic radiation on living beings.

Sputnik X 1961 Iota		Mar. 24, 1961 64°54' to equator	241.3	178.6	88.4	456

Further testing of construction of the space ship and systems to insure necessary conditions for manned flight. Radio tracking and communications systems; TV apparatus to relay information on condition of animals on board.

Vostok I 1961 Mu		Apr. 12, 1961 65°41' to equator	302	188	89.0	4,725

Placed manned space ship in earth orbit and recovered; included radio and TV equipment relaying information on condition of man on board; life-support equipment.

Vostok II		Aug. 6, 1961 64°56'	257	178	88.6	4,731

Placed manned spaceship in earth orbit and recovered.

the Antarctic, which are equipped with intensity monitors of standard design. The data obtained are being used to study the mechanism of cosmic ray modulation in interplanetary space, with special emphasis on analysis of the superposition effects of solar plasma clouds and the trapping of solar particles in these clouds. In the near future, Argentina plans to augment these activities with a program of similar measurements at high altitudes, first with balloons and later with sounding rockets. Lying in the low geomagnetic latitudes of the southern hemisphere, Argentina can provide optimum location for high altitude measurements that will overcome some present difficulties in obtaining accurate measurements of albedo neutron flux.

Argentine scientists have been operating, under a contract with the United States Air Force, a network of three optical tracking stations equipped with ballistic cameras. The Astronomical Observatory of San Juan is now arranging to purchase the equipment involved so that this work may be continued. Another optical tracking station, located at Villa Delores is operated and equipped with a Super-Schmidt camera, through an arrangement with the Smithsonian Institution. Observations from these tracking stations will be used for geodetic research. Argentina has a volunteer moonwatch team located at San Juan.

In conjunction with the high altitude cosmic ray program, Argentine scientists are beginning research in related sciences, including studies of the ionosphere, aurora, geomagnetism, and solar physics. In November, 1960, Argentina organized a symposium on space research in Buenos Aires which attracted scientists from all over Latin America and from the United States and Europe.

AUSTRALIA

Australia continues to co-operate with the United States in tracking and in receiving telemetry from satellites, and with the United Kingdom in rocket research, and Australian scientists conduct investigations of their own based on data taken from passes of satellites over Australia and from rocket firings at Woomera.

Tracking and telemetry activities include operation of a station in the United States Minitrack network, a station in the United States Mercury network, a Baker-Nunn camera station, and a deep space instrumentation facility, all of them located at the space research station at Woomera. An additional Mercury station is located at Perth. The Minitrack equipment, in operation for just over three years, has been used to make about 5,000 records of satellite passes. About the same number of telemetry recordings have also been produced, along with about 150 Doppler frequency measurements. The Baker-Nunn camera was put into operation in March, 1958, and since that time 3,300 satellite passes have been recorded and some significant range records have been achieved, notably, photographs of the 6-inch diameter Vanguard I satellite at a distance of 2,500 miles.

The deep space instrument facilitiy, located at Woomera, is part of the three-station network being established by the Jet Propulsion Laboratory under contract to the National Aeronautics and Space Administration (NASA). The other two stations are Goldstone, California, and Johannesburg, South Africa. The three stations are approximately 120° of longitude apart. The Woomera station, which has been in operation since November, 1960, consists basically of an equatorially mounted 85-foot diameter radio telescope with ancillary control equipment, along with electrical equipment for receiving, recording, and converting telemetered data. To date, this facility has been used for a moon-bounce experiment with the station at Goldstone and for tracking the Explorer X magnetic field probe. Plans call for the addition of low noise pre-amplifiers to extend the recording range of this radio telescope from the present limit of 7,000,-000 miles to several thousand million miles for a probe carrying a 5-watt transmitter. It will also be used for radio astronomy observations.

Woomera continues to be used as the ground range for the launching of experiments aboard the British Skylark rocket, in a co-operative venture with the British Royal Society and the United Kingdom Ministry of Aviation. Three such rockets were launched during the IGY and another seven during 1959. Experiments carried by the last seven included ion probes, ion mass spectrometers, solar X-ray cameras, a solar Lyman-alpha counter, and sodium vapor trails and grenades for the study of temperature and winds. These experiments will be discussed in more detail as part of the program of the United Kingdom. In a United States, United Kingdom, and Australian co-operative project to survey ultraviolet radiation in the southern sky, instruments developed by NASA are being launched aboard British Skylark rockets from Woomera.

Australian scientists have undertaken the analysis of signal recordings from satellites for use in studies of variations in electron concentrations in the ionosphere. By comparing these data with records from standard ground-based ionosonde equipment, much useful information has been obtained on changes in ionization densities above the F_2 layer maximum.

Studies of the height distribution of airglow intensities will be made by using sounding rockets to measure infrared emissions. To study upper atmosphere and high altitude winds, artificial clouds are being produced by exploding grenades ejected from Skylark rockets at heights of 100 to 150 kilometers. The clouds appear blue-white to the eyes and persist long enough to be photographed.

A direction-finding radio receiver is being used to detect the strength and direction of very low frequency radio noise in the atmosphere. These data are useful in the study of ion density and will provide valuable comparisons with similar observations from satellites. Upper atmosphere density is being measured by use of special gas gauges aboard Skylark rockets.

The work of five moonwatch groups has been particularly valuable in view of the scarcity of good observations from the southern hemisphere. Among their notable achievements were the first sightings of Sputniks I and II, and the first sightings of Vanguard I and Explorer IV when the latter satellites were the subjects of world-wide searches.

BELGIUM

Belgium has organized a national space committee and a national center of space research; both are staffed by scientists from

Space Programs of Other Nations

universities and scientific institutions. Their present program includes visual observation of satellites, and theoretical studies of satellite orbits and of the structure of the upper atmosphere. The latter studies have been of value in the development of a standard reference table (COSPAR International Reference Atmosphere) describing upper atmosphere parameters particularly for altitudes above 32 kilometers. This, in turn, is related to intensive theoretical studies of the mechanisms that produce the observed temperature and density patterns of the upper atmosphere.

CANADA

Because of special scientific opportunities of Canada's geographical position relative to the earth's magnetic pole, Canadian scientists are concentrating their efforts on near space—between about 35 and 300 kilometers, or the range between balloons and satellite altitudes—and to some extent on objectives related to the interaction of solar radiations and particles with the earth's magnetic field. Canadian scientists co-operating with United States rocket experimenters using the facilities at Fort Churchill, have launched two experiments aboard Nike-Cajun rockets, and two aboard Aerobee-type rockets. These experiments measured electron densities and cosmic rays. Canada also has been developing its own solid propellant rocket, the Black Brant. Five Black Brants have been launched carrying experiments to measure cosmic rays, to study auroral absorption, and to study radiation effects associated with seeding the upper atmosphere with nitric oxide.

Along with efforts to measure electron and particle densities in the ionosphere directly, the Canadian program also includes a number of ground-based studies of solar terrestrial relationships; these include ionosphere soundings, riometer measurements, auroral observations, and solar and galactic radio noise measurements. Particles and radiations responsible for radio absorption in the lower ionosphere also are being studied.

Canadian scientists are preparing an instrumented satellite, the "topside sounder," to be launched by NASA in 1962. This experiment is, in effect, an upside down vertical-incidence ionosphere sounder that makes pulse soundings from above the ionosphere rather than from below. The same satellite also will carry cosmic ray particle detectors of five different types. Three new satellite recording stations are being established for this Canadian satellite experiment, at Ottawa, Prince Albert, and Resolute. A Minitrack station has been installed at St. Johns, Newfoundland, as part of the NASA chain of tracking stations.

CZECHOSLOVAKIA

Czechoslovakia has been active in optical tracking of satellites and in theoretical studies. Optical tracking stations at Praha, Brno, Bratislava, Pezinck, and Skalnaté Pleso use special telescopes, cameras, and star maps. Observations have been made of about 100 satellite passes, including passes of Tiros I, Echo I, Sputnik IV, and Sputnik V. Theoretical analyses are being undertaken of the impact of Lunik II on the moon. Satellite orbital data are used for studies of periodic and secular orbital variations, the influence on satellite orbits of the ellipicity of the earth's equator, the determination of parameters of the earth's field of gravity, the perturbations of a stationary satellite and the stability problem involved in satellite motion around two of the equilibrium points in the earth-moon system, and studies of the structure of the upper atmosphere.

FRANCE

French scientists have been active in rocket research, satellite observations, and theoretical studies. A series of successful test launches of the Veronique rocket was carried out in 1960, and new rockets are being developed to carry payloads to still higher altitudes. A number of successful scientific experiments have been conducted using the Veronique rocket. From four rockets, reagents were injected into the upper atmosphere forming alkaline clouds—one of sodium, two of sodium and potassium, and one of lithium—for studies of upper atmosphere winds. A 65-kilogram charge of dynamite was exploded to disperse the reagents at an altitude of 156 kilometers creating a luminous cloud

15 kilometers in diameter which was the object of acoustical, optical, magnetic, and electron measurements. Turbulences were observed beneath 100 kilometers; great variations were discovered in the density of the atmosphere between 150 and 200 kilometers, and temperatures were correctly determined for the first time at altitudes between 130 and 150 kilometers. By determining the coefficient of diffusion of these alkaline clouds to a distance of 190 kilometers, variations in atmospheric density were calculated; the concentration of atomic oxygen was deduced from the reaction time between the clouds and the atmosphere.

In another series of five rocket experiments, the diffusion of alkaline clouds was used to study the sodium emission spectrum. The first rocket fired at dusk allowed observations of disturbances of the green spectral line against the night sky for three hours. Atmospheric pressure, the radiation field of radio transmitters, and the atmospherically scattered sunlight and airglow were measured. Correlation of these data provided information on pressure, density, and electron concentration as a function of height and time of day. In another experiment, a rat was carried to an altitude of 106 kilometers in an aluminum capsule inside the nose cone of a Veronique rocket. Action potential measurements were made of the cerebral cortex and of the mesencephalic network, the muscles of the neck, and the anterior arches of the diaphragm, and the rates of heartbeat and respiration were recorded. Measurements also were made of the interior noise, temperatures, and pressures, and of the acceleration stress on the capsule. French scientists also have been active in solar research and in studies of interstellar materials, particularly the nebulosities around hot stars observed by means of rocket flights.

Three French observatories (Meudon, Besançon, and Strasbourg) and a network of fifteen weather stations were active in visual observations of satellites, using theodolites and chronographs. These observations are for orbital studies and other purposes of interest to astronomers (such as celestial mechanics) and geodesists. Photographs of the Echo satellite were taken at Meudon and Strasbourg. In addition, satellite radio telemetry has been monitored for Doppler shift studies aimed at improving position forecasts and for the study of wave propagation characteristics of the ionosphere. France also participated in an Echo I experiment by receiving signals bounced off the Echo satellite from Holmdel, New Jersey.

GERMANY

German scientists have been carrying out theoretical studies based on data from the motions and radio transmissions of satellites, with a view to their usefulness in investigations of the physical properties of the ionosphere and upper atmosphere, as well as low-level studies of cosmic rays and X-radiation from auroras. The propagation and intensity of the 20 megacycle Sputnik III signals were analyzed for information about three kinds of variations in the F_2 layer of the ionosphere—day-night, 27-day solar rotation period, and annual cycles above the 300-kilometer level. These same data, along with data on satellite tracking, were used to develop a theoretical model of the ionosphere accounting for these variations. Interpretive studies are going forward on the way in which radio wave propagation from satellites is affected by the height and density of the ionosphere. In these studies, the ray paths of radio emissions from the satellites are calculated for large angles from the zenith, and these calculations are being checked by studying the field intensity of radio waves received by rockets from transmitters on the ground. Riometers, echo soundings, Faraday rotation, and back scatter in satellite wakes also are being used to study the ionosphere.

Another theoretical study by German scientists provides a discussion of correlations that do, or may, exist between atmospheric drag variations in satellite motions and three naturally occurring phenomena, some of which have been observed for some decades: auroral activity, especially if the satellite passes through the auroral zones; upper atmospheric lunar tides; and 27-day solar rotation periods. When these studies are followed by experiments and observations, they are expected to yield information on changes in the density of the atmosphere as affected by these three naturally occurring phenomena. Theoretical interpretation is being made of the connection between the time variations of the cosmic ray energy spectrum and the electrical potential of the earth as both are affected by the variations in solar activity. The time pattern of the arrival of particles of different energies from the sun has been measured and inferences drawn about the effect of the earth's magnetic field in delaying the arrival of particles of different energies. X-ray bursts originating in the auroral zone are being observed with instruments flown by balloons from northern Sweden. Results from United States rocket launchings are being analyzed to yield information on the relationship of atmospheric heating to circulation above 55 kilometers and to seasonal reversals of high altitude wind patterns.

German scientists are active in ground-based space research. For example, they are monitoring solar radio frequency bursts and making systematic observations of airglow. Some are working on the theory of variations produced by the earth's magnetic field in corpuscular solar radiation, and others are studying the interaction of this radiation with comet tails as revealed by spectrograms. A theoretical model has been developed to explain zodiacal light in terms of scattering and polarization by mixtures of dust particles of various sizes.

INDIA

Indian scientists have been active in precision optical tracking of satellites, reception of radio signals from satellites, meteorological studies and associated research concerned with the ionosphere, studies on the upper atmosphere and radio astronomy. Since June, 1956, the Uttar Pradesh State Observatory has cooperated with the Smithsonian Astrophysical Observatory in the operation of a Baker-Nunn camera. This camera fills an important gap between Iran and Japan in the world network of twelve of these tracking cameras. Approximate positions of both American and Russian satellites are obtained with respect to fixed reference stars and are fed into the world rapid communication network for information on satellites. In addition, radio transmissions from satellites are received, including time of transit, duration and amplitude of signals, and Doppler shifts in frequency.

Meteorologists in India have been particularly interested in the meteorological satellite program of the United States and Tiros I cloud cover pictures over India have been studied and analyzed. Arrangements have been made to correlate radiosonde, rawinsonde, and ground-based cloud cover observations with passes of future Tiros-type satellites.

Data from United States and Soviet satellites have been used by scientists at the National Physical Laboratory in New Delhi to develop a model of the earth's atmosphere, and records of cosmic radio noise are being used in conjunction with ionograms and automatic ionosondes to obtain estimates of ionization in the outer ionosphere. Equipment is being installed at the Physical Research Laboratory in Ahmedabad to receive telemetry from the Explorer XI gamma ray astronomical satellite.

ISRAEL

On July 5, 1961, Israel launched a rocket known as the Shavit II, which discharged a metallic sodium cloud at the height of 80 kilometers to study wind conditions in the ionosphere.

ITALY

Italy organized its first national space research program in 1960. Initial efforts were directed at building up and improving ground-based research facilities, including optical and radio tracking, and preparation for Italy's first launchings of instrumented rockets. Ground-based research is being developed at Italian universities in Florence, Bologna, Rome, and Asiago. A new laboratory has been built at the University of Rome for research in the problems of re-entry into the atmosphere. This laboratory will include a space simulator for studies of scientific instrumentation to be placed aboard satellites and space probes.

In co-operation with NASA, Italy launched its first scientific rocket, a Nike-Cajun, in early 1961 from the rocket range in Sardinia. The rocket released a sodium-lithium cloud at an altitude of 160 kilometers for studies of the dynamics and structure of the upper atmosphere. Observations were conducted from a network of seven stations, five in Sardinia and two in continental Italy. A second series of sodium-lithium cloud experiments was ejected from rockets launched from Sardinia in April, 1961, in co-ordination with similar experiments launched by NASA from Wallops Island, Virginia. Instruments are now being developed to measure cosmic rays and geomagnetic fields at rocket altitudes.

Signals from satellites are being used by Italian scientists to study propagation characteristics of the ionosphere, and systematic observations of satellites by the astronomical observatory at Asiago have obtained good results for use in determining seventy-nine positions for satellites, including Echo I, Sputnik IV, and Explorer VII. These observations are used for computations of satellite co-ordinates relative to the Asiago station and of orbital elements. A new method is being developed of tracking satellites based on the simultaneous reception of signals at four separate stations. Methods for determining the distribution of meteors based on radar echoes from meteor trails are being investigated. Other investigations of meteor density are based on photographic techniques. A transistorized apparatus for the reception of satellite telemetry is being developed.

JAPAN

Japan has continued its development of the Kappa sounding rocket and has been developing new techniques for the instrumentation of these rockets. Ten Kappa rockets have been launched since the end of the IGY period, four of them in 1959. Two of these were completely successful launches employing the grenade method to measure atmospheric temperature and wind. Rockets launched in 1960–61 measured temperature and wind velocity up to an altitude of about 70 kilometers, and ion density and cosmic rays up to 200 kilometers. One rocket flown in 1961 measured the electron temperatures and airglow at about 170 kilometers.

Japan launched its first three-stage rocket to an altitude of 350 kilometers in April, 1961. This new rocket will be used to measure temperature and wind velocity, electron and ion densities, electron temperatures, airglow, radio noise, the magnetic field, and cosmic rays.

Radio observations of satellites are made by the Radio Research Laboratories at Kokubunji. Measurements of the Doppler shift in frequency and in radio field intensity were made for all passes of Sputniks III, IV, V, and VI and precise variations of the field intensity were analyzed. Telemetry signals from Explorer VII were recorded on magnetic tape for all passes from October 14, 1959, until October 28, 1960, and Japanese scientists have proceeded with their own analysis of these data. The Radio Research Laboratories now has under construction a rotatable parabolic mirror antenna with a diameter of 30 meters for tracking and receiving data from deep space probes, and for work with communications satellite experiments.

A Baker-Nunn camera has been in operation at the Tokyo Astronomical Observatory as part of the twelve camera worldwide network of the Smithsonian Astrophysical Observatory. Japanese Moonwatch teams have made a very large number of successful observations.

THE NETHERLANDS

Scientists in the Netherlands have made regular visual observations of satellites and have been receiving telemetry satellite signals ever since the launching of Sputnik I. These observations are reported to the observatory at Utrecht and are relayed to the United States and the Soviet Union. Signals from the United States satellite Solar Radiation III are being received and analyzed at Utrecht for studies of solar Lyman-alpha and X-radiation. Dutch scientists also have been active in space-related studies in astronomy, radio astronomy, in ionospheric, cosmic ray and plasma physics, and other fields. In particular, studies of the solar corona, of interplanetary matter, and of ionospheric soundings are directly related to experiments conducted in space. A Baker-Nunn camera is located at Curaçao, in the Netherlands West Indies.

NORWAY

In 1962, under the aegis of the Norwegian National Space Committee, Nike-Cajun rockets will be launched from Andoya carrying instruments developed in co-operation with the Technical University of Denmark at Copenhagen. Some technical consultation and training will be provided by NASA. There are plans to study the auroral ionosphere, particularly the altitude, density, and average collision frequency of electrons in the lower layer of the ionosphere.

PAKISTAN

The Pakistan Upper Atmosphere and Space Research Committee plans to launch rockets in 1962 which will eject sodium-lithium clouds for studies of turbulence and wind patterns in the upper atmosphere. Eventually the committee plans to co-ordinate such observations with similar activities at other locations, such as Wallops Island in the United States and Sardinia in Italy. Two Pakistani graduate students will attend United States universities active in space research work under a NASA fellowship program, and several Pakistani engineers are learning the requirements of the scientific sounding rocket program at NASA installations.

PERU

Space research in Peru is conducted in co-operation with NASA and with universities in the United States. A Minitrack station and a telemetry receiving station are located in Lima. Most of the technical staff of these stations is provided by the Huancayo Geophysical Institute, and two Peruvian scientists are participating in an advanced training program at the NASA Goddard Space Flight Center. A Baker-Nunn camera is operated at Arequipa in co-operation with the Arequipa University and the Huancayo Geophysical Institute, and photographic tracking stations are located at Huancayo, Talara, and Chosica. A telemetry station is being installed at Huancayo as part of a Pennsylvania State University research program associated with NASA's proposed ionospheric beacon satellite.

A large radar scatter observatory nearing completion near Lima will be capable of measuring upper atmosphere parameters to a height of 3,000 kilometers and will provide a basis for comparison with data retrieved from satellites on magnetic fields, temperatures, and electron densities. This observatory is being built by the United States National Bureau of Standards in co-operation with the Huancayo Institute. The Huancayo Institute also is making ground measurements of the magnetic field, the ionosphere, cosmic radiation, solar phenomena, riometer absorption, and gravity.

POLAND

Polish scientists have been active in visual and photographic observations of satellites and rockets in close co-operation with the space research program of the Soviet Union. Observations made by about ten stations are cabled to Cosmos in Moscow. Radio observations of the sun and moon are being made at observatories at Kracow and Torun. Polish scientists also have been active in ground investigations related to space research including cosmic ray and plasma physics, and in theoretical research in the magneto-hydrodynamics of the interstellar medium.

SWEDEN

The International Meteorological Institute of the University of Stockholm conducted its initial sounding rocket experiment in 1961. An Arcas rocket supplied by the United States was launched from Kiruna. This is the first in a series of experiments aimed at studying noctilucent clouds, and plans are being made for more extensive studies of the upper atmosphere and ionosphere.

UNION OF SOUTH AFRICA

Scientists in South Africa are tracking satellites and recording satellite telemetry data. A radio tracking station, part of the United States network, is located at Esselen Park some twenty miles northeast of Johannesburg. It was established during the IGY and has continued in full-time operation since then. An optical tracking station, located at Olifantsfontein, Transvaal, is part of the network of twelve Baker-Nunn cameras established during the IGY by the Smithsonian Astrophysical Observatory. In co-operation with the Jet Propulsion Laboratory, a deep space tracking facility is being established at Hartebeesthoek some forty miles northwest of Johannesburg. Its main piece of equipment will be an 85-foot steerable antenna.

Moonwatch teams are located in Pretoria, Cape Town, and Bloemfontein, South Africa, and in Bancroft, Northern Rhodesia. Notable accomplishments of these teams were the first observation of Explorer IX and of the 12-foot sphere satellite. Meteorologists from the South African Weather Bureau are planning to co-operate in world-wide meteorological observations in connection with United States Tiros satellites.

UNITED KINGDOM

Scientists in the United Kingdom participate actively in satellite tracking and data recovery, in experiments with vertical sounding rockets, and in theoretical satellite studies, and have several programs for the development of new space research instruments and techniques. The 250-foot radio telescope at Jodrell Bank tracks deep space probes, and has received Pioneer V signals from the record distance of 22.5 million miles. The same radio telescope was used for communications experiments with Holmdel, New Jersey, via the Echo satellite.

A United States Minitrack tracking station at the Radio Research Station in Slough is part of the world-wide NASA network for receiving and recording data telemetered by satellites. Optical tracking stations are operated at the Royal Observatory in Edinburgh, and by the British Meteorological Office on the island of Malta. Data from these stations are fed into the Spacewarn satellite prediction service to assist in the determination of satellite orbits, and the prediction of satellite locations for tracking stations in other parts of the world. In addition, British scientists are working on new designs for optical tracking in order to overcome some of the present limitations in accuracy and sensitivity, and the high cost of establishing and operating such stations. Photoelectric techniques and image intensifiers are being explored to extend the threshold of sensitivity beyond that of the human eye and of photographic emulsions. Other forms of special cameras are being developed at the Royal Radar Establishment.

British scientists are conducting experiments to study the structure and composition of the atmosphere, radio propagation in the ionosphere, geomagnetism, cosmic radiation, and electromagnetic radiation. Atmospheric temperature, pressure, density, and winds to the height of 90 kilometers are studied by the use of luminous glows following the explosion of grenades ejected from rockets. The changing shape and size of a luminous glow permits determination of diffusion coefficients and local winds. It is believed that the luminosity is a photochemical effect arising from reaction with atomic oxygen. Plans are being made to extend the altitude of this type of study by tracking light balloons ejected from rockets, and further extension based on the grenade method is planned to provide a systematic investigation of diurnal and seasonal variations. Individual measurements of temperature, winds, and diffusion rates are being made by other British scientists, who use interferometers to photograph the interference fringes of sodium trails ejected from rockets. A group at Belfast, Ireland, is studying airglow characteristics and plans to measure airglow heights by launching Skylark rockets equipped with photometers sensitive to various airglow radiations.

A substantial part of the British space research effort is concerned with the electrical state of the atmosphere and its relation to the type of radiation believed to be mainly responsible for the ionosphere. Skylark rockets are being used to measure electron density and positive ion content of the ionosphere. As a logical extension in height of the rocket studies, British scientists are preparing instrumentation for a satellite to be launched by NASA to measure electron density, electron temperature, and positive ion temperature. Other equipment aboard this satellite will measure solar X-rays and ultraviolet radiations; correlation of these observations is expected to be valuable in elucidating the physics of the ionosphere.

Also in progress are determinations of air density based on analyses of satellite orbits and on studies of air mass density variations under various conditions, such as day-night differences, and density differences between periods of maximum and minimum sunspot activity. Upper atmosphere winds are being studied by radar tracking of bundles of resonant dipoles released from Skylark rockets in Australia. Small rockets will be used to raise the present 30-kilometer radiosonde ceiling to 60 kilometers, and eventually to 100 kilometers. The vertical distribution of ozone in the atmosphere is under study by rocket measurements of solar radiation absorption in Australia, and plans are being developed for mounting similar instruments in a future satellite. An analysis is being made of the possibility of using a satellite to measure total terrestrial radiation leaving the earth and reflected solar radiation. Spectral measurements of the atmospheric infrared continuum have been made as an aid to the interpretation of the results of this program.

Other groups of British scientists are studying the ionosphere using radio propagation from rockets and satellites. Instruments to measure Doppler frequency shifts have been launched aboard Skylark rockets and plans are proceeding for a pulse-propagation experiment. Another experiment, in which the nature of the wave field above a long-wave transmitter is explored by rocket instrumentation, is intended to yield deductions about the electron distribution in the D region of the ionosphere.

At Jodrell Bank, ionospheric studies are under way using the transmissions from several satellites, particularly Vanguard I, and irregularities of several per cent in the total ionization of the atmosphere have been observed. A study of scintillations in Sputnik III transmissions has been undertaken jointly with scientists at other European observation stations. Ionospheric information of great interest is expected from analysis of radio signals received simultaneously at three widely separated stations.

British geophysicists are studying rocket and satellite techniques for exploring the geomagnetic field, and are planning to launch a proton precession magnetometer aboard a Skylark rocket.

British geodesists are using satellite observational data for studies of orbital theory. One theory being developed provides analytical expressions for the effect of aerodynamic forces on orbits in an oblate rotating atmosphere. A second theory has been formulated to provide for the effects of solar and lunar perturbations, including solar radiation pressure, on satellite orbits. These and earlier theories have been applied to the observed orbits of satellites in determinations of the earth's gravitational field, and new and better values have been obtained for the coefficients of the second, fourth, and sixth harmonics in the earth's gravitational potential. Other studies are under way, particularly for those satellite orbits in which the mean motion bears a simple ratio to the angular velocity of the earth's spin.

Micrometeorites are being studied by use of Skylark rockets carrying microphone detectors and thin aluminum foils. Holes formed in the aluminum by impacting particles let in sunlight and the transmitted light pulses are detected by silicon cells. Calibration of the results permits an estimation of micrometeorite size.

Other investigations concern the interrelationship between cosmic ray intensity, solar activity, and the structure of the interplanetary magnetic field. Further theoretical work is being carried out on the behavior of primary cosmic rays in the geomagnetic field and, using scale model experiments, on the effects of various field configurations, of both internal and external origin, on cosmic ray threshold rigidities. The first satellite instrumented by British scientists will include a cosmic ray detector capable of measuring both the total charged particle flux and the flux of heavy nuclei. Instrumentation is being developed for both rocket and satellite investigations of the higher energy components of primary cosmic rays, in particular to determine whether there is an uncharged component and, if there is, its magnitude and any anisotropy in its direction of incidence.

Equipment is being designed for a study of galactic electromagnetic radiations down to frequencies of about one megacycle. This instrument will also permit determinations of electron density in the upper ionosphere. At Jodrell Bank a detailed analysis of ionospheric effects on radio waves at frequencies of the order of one megacycle traveling from a distant point in space to a satellite has indicated the presence of focusing interference effects in the ray paths. A satellite experiment has been proposed to determine the spatial disturbance of radiation at these frequencies by using the occultation of the earth.

Measurements have been made of solar X-ray emissions as part of the program of rocket and satellite measurements of X-ray intensities. Plans are well advanced for direct measurements of the solar spectrum using a proportional counter X-ray spectrometer carried in a Skylark rocket. These measurements will be used in studies of the variations of spectral shape and intensity with solar activity over a period of time. Solar X-ray measurements also will be made in the first British-instrumented satellite by a proportional counter with a very thin beryllium window. Plans are well advanced for observations to determine stellar magnitude and for a survey of ultraviolet radiations in the southern hemisphere. Instrumented packages developed in cooperation with NASA will be flown aboard Skylark rockets from the Woomera range in Australia. Design studies are advancing for a satellite to make astronomical observations in the ultraviolet.

OTHER WORK

Many more countries are assisting the general space effort, particularly in tracking satellites. Radio tracking stations of the United States Minitrack network are located at Guaymas, Sonora, Mexico; at Quito, Ecuador; at Santiago, Chile; and in Bermuda. Optical tracking stations of the Baker-Nunn network are located at San Fernando, Spain; at Shiraz, Iran; and at Santiago, Chile. Amateur tracking teams of the Moonwatch system not mentioned above are active in Brazil, Canada, Chile, Germany, Mexico, the Philippines, Taiwan, the United Kingdom, and Uruguay. About thirty countries are co-operating in ground-based weather observations co-ordinated with passes of Tiros-type satellites.

A sufficient number of common threads runs through the various national programs to indicate fruitful areas of collaborative effort. For example, as meterological observations at rocket altitudes become available systematically over increasingly large geographic areas of the world, extremely useful correlations will be possible. A further step in this collaborative planning would be the launching of such meteorological rockets during the same time intervals to obtain synoptic observations over increasingly greater areas of the world. Later, similar co-ordinated measurements could be made at high altitudes for studies on the upper atmosphere that require larger rockets. Such larger rockets will continue to be the essential tools for upper atmosphere investigations at altitudes below satellite perigees. Much work can be done in programs based on balloon-borne experiments, complementing similar work carried out at high altitudes with rockets and satellites. Moreover, there are numerous opportunities for scientists everywhere to engage in other types of space research just as investigations in many phases of geophysics have been intensified and expanded as a result of the IGY and the advent of the space age.

PART FOUR

International Space Co-operation

Introduction
International Space Organizations
International Programs of NASA
The United Nations and Outer Space
COSPAR and Space Co-operation

INTRODUCTION

HUGH ODISHAW

Perhaps no development in history except space exploration calls so clearly for co-operation among men and nations. Space activities not only transcend national boundaries but reach beyond earthly ones. Once launched into space, an earth satellite becomes a cosmic body like the earth's natural satellite, the moon.

The four essays in this section address themselves to the topic of international co-operation in space activities. The first chapter outlines the space organizations presently active, for the advent of the space age is so recent and its development so rapid that the existence and work of many of these organizations remain unknown, even to those who seek to encourage co-operation. Three other essays discuss multi-national co-operative space programs conducted by the United States; the functions of the United Nations and its pertinent specialized agencies; and co-operation in the world's scientific community under the auspices of the International Council of Scientific Unions.

These essays provide information useful for examining the problems and prospects of co-operation in space. There are many areas where co-operation is desirable: adventure and exploration, practical applications, military applications, research in almost every field of science, and even efforts to control our environment. Mechanisms and procedures for co-operation in one field may not be suitable in another. The problems associated with space co-operation are divided into two categories: regulation and control, and collaboration in research.

Regulation and control are political in nature, and, accordingly, require governmental agreements. This indicates that they fall within the scope of the United Nations. The most critical problems of control—avoiding war, maintaining peace—are now linked to space because vehicles in near space can assume military functions. This is a new prospect and problem; it further complicates past deliberations relating to nuclear testing and disarmament. The solution of this problem will not be easy.

Regulatory problems also require intergovernmental agreements. The International Telecommunications Union, one of the specialized agencies of the United Nations, has dealt with the allocation of radio frequencies for many years, and has recently turned to the communications problems and needs of space systems. Another United Nations agency, the World Meteorological Organization, is studying weather applications of satellites. But much more remains to be defined and undertaken. Prior to the last decade or so, man could destroy his near enemy, pollute his own environment, or even by abuse lay waste local land masses for centuries; but he could not affect the total human environment. Aside from the threat of nuclear weapons in warfare, the detonation of these devices for test purposes produces radioactive fallout that spreads around the world. Whether the level is or will be dangerous is beside the point here; it is the fact that an effect is global which generates one aspect of concern over testing. Space programs provide vehicles which will permit man to influence his total environment in a number of ways. Injecting substances into the upper atmosphere influences the earth's magnetic field and the ionosphere, and weather and climate may be controlled by satellite carriers. The potential range of such acts may thus call for the establishment of new international regulatory bodies under the auspices of the United Nations.

In scientific research an appreciable tradition of co-operation exists. Scientists have increased their international activities since the last war, and governments and intergovernmental bodies have shown increased interest in scientific co-operation. The United Nations and its Committee on Peaceful Uses of Outer Space have enunciated their interest in space, and have recognized the role of such non-governmental organizations as the International Council of Scientific Unions and its Committee on Space Research (COSPAR). Such interest is inevitable as the impact of science upon human affairs increases, but it may also reflect realization of the general value of scientific co-opera-

Introduction

tion, as demonstrated by the International Geophysical Year. The success of this large endeavor, apparently attained with ease by the scientific community, yielded not only the scientific results that had been sought; it also engendered good will and considerable public interest, and established a promising climate for further, similar ventures. Moreover, the climate made possible at least one major political event which grew out of the International Geophysical Year—the Antarctic Treaty. This is the only treaty in existence consecrating a portion of the earth to peaceful purposes, and it serves as a hopeful precedent.

The space age itself was initiated by the International Geophysical Year, which established co-operation that was fortunate in several ways. First, it made the results of the International Geophysical Year space program available to all. Second, it served to minimize what might have been an even more competitive situation between the two satellite-launching nations. Third, it provided the foundation for its successor in space research co-operation—the Committee on Space Research, created after the International Geophysical Year.

COSPAR has effectively continued the work of the International Geophysical Year, extending the co-operative aspects of space research. But COSPAR has not as yet carried out a major space effort comparable to that of the International Geophysical Year. Such an effort would be an important step, not only for science, but because its scientific success would also serve as a contribution to peaceful, productive pursuits. COSPAR has undertaken the rocket and satellite program for the International Year of the Quiet Sun (1964–65), which will be devoted to investigations of the upper atmosphere, the interplanetary medium, and solar and terrestrial relationships during the coming sunspot minimum. This program is important for the Year of the Quiet Sun, and lays a foundation for COSPAR's formulation of subsequent and more ambitious space efforts.

In formulating a major space effort, the principal problem before the scientific community has been the costs and limited availability of space tools. This situation is changing for several reasons. First, the tools for space research are becoming more available, and their availability will continue to increase, probably at an accelerated rate. During the International Geophysical Year only a few nations had rockets; the number has doubled, partly because the United States has made sounding rockets available to several countries, and partly because a few more countries have undertaken their own development of rockets. Although the United States and the Soviet Union are still the only powers which have launched satellites, this is destined to change within the near future, as other nations or groups of nations (such as those in Western Europe) arrange joint launching enterprises. Moreover, the United States has agreed to launch satellites and payloads prepared abroad under the auspices of COSPAR; the first of these, a United Kingdom satellite, was launched in April, 1962.

Second, the results of space research have made themselves felt. It is now obvious that the significant advances in research in many fields will be made by observations and measurements conducted aboard spacecraft. Appreciation of this fact has intensified scientific interest throughout the world, directing attention to all facets of space research, for much can be done without direct access to spacecraft. For example, the data from space experiments can be used for analysis and theoretical research by scientists everywhere. Some satellites broadcast data continuously. Data can be received by all scientists, and analyzed and interpreted as if the experiment were their own. Most satellites record and transmit data on command to specific receiving stations, but as the problems of telemetry and power supply are solved, it is possible that more experiments could be planned with continuous transmission of measurements.

Moreover, a large area of research—investigations from the ground, carried out with the conventional tools of geophysics and astronomy—now acquires fresh and even pressing interest. Astronomy during this century has been mainly concerned with galactic and extra-galactic topics; our own solar system—with the partial exception of the sun itself—has concerned few astronomers. Yet much can be learned with conventional techniques about the moon and planets, about the particles, radiations, and magnetic fields in interplanetary space, and about the sun. Such knowledge is a legitimate and worthy scientific goal, but

Introduction

it is also related to the needs of future space enterprises. If spacecraft, manned or unmanned, are to voyage to other bodies in the solar system, the more we can learn about the interplanetary medium, about the nature of lunar and planetary atmospheres, ionospheres, magnetic fields, and about their geography and geology, the more intelligently such ventures can be designed. Whatever can be learned from ground-based experiments and from theoretical research reduces the demand upon costly space carriers, leaving opportunities for those investigations which can only be conducted from spacecraft. In short, there are many opportunities for scientific participation in space research throughout the world.

For all international activities, whether requiring formal governmental arrangements or pursued somewhat informally by the scientific community, co-operation is crucial. If a problem involving regulation is to be resolved in a political arena such as the United Nations, there must be governmental willingness to agree to the regulatory provisions that may be adopted, and to co-operate in sustained support of these provisions. At the other extreme—co-operation among scientists—governments must at least acquiesce. Thus the attitudes of governments can affect every area of space activity.

Each space activity, real or potential, calls for analysis. Topics requiring the forum and power of intergovernmental bodies need to be distinguished from those best served by allowing scientific bodies to act. Success in all areas requires that nations be willing to set aside, at least to some extent, national prerogatives. It is likely that political and national considerations will make questions that are a proper function of the United Nations difficult to solve. If this continues to be so, then the need for co-operation in science becomes even more pressing, for such co-operation could lead not only to a sounder and more economical program of space exploration and research but also to a background of co-operative acts, events, and procedures upon which more might be built. This line of reasoning implies that a great responsibility rests upon the scientific community, particularly as represented in COSPAR, to formulate imaginative, sound, world-wide programs of space investigation, comparable in breadth and spirit to the International Geophysical Year. The dimensions of the space age merit no less than this, for the opportunities and problems of space are unparalleled even when viewed from a narrowly physical point of view. Until now the total content of the universe directly accessible to mankind has been the earth itself and a small portion of the atmosphere about it; now the accessible content has been multiplied suddenly more than a million times.

INTERNATIONAL SPACE ORGANIZATIONS

LEONARD E. SCHWARTZ

The wide-ranging impact of outer space is dramatically revealed by the many organizations, both governmental and non-governmental, that have expressed an interest in space affairs. Thus far the more prominent and substantive international space role has been performed by the non-governmental agencies such as the International Council of Scientific Unions (ICSU) and its various member unions concerned with theoretical and applied mechanics, geodesy and geophysics, astronomy, pure and applied physics, and others.

As the investigation of outer space became a factor of political concern, governmental organizations such as the United Nations, through its special committees on space, and the specialized agencies such as the International Telecommunications Union (ITU) and the World Meteorological Organization (WMO) and the United Nations Educational, Scientific, and Cultural Organization (UNESCO), expressed a heightened interest in the development and control of space activities. (For space activities of the United Nations and specialized agencies, see pages 277–90.)

Indicative of the growing interest in space is the formation of regional space organizations, especially in Western Europe and, to a lesser extent, in Latin America. These are being set up because of an awareness that capabilities, skills, and costs exceed the resources of the individual nations within these regions, and indeed, unless there is co-operation, it is questionable whether they can develop an effective and meaningful space program. The European Space Research Organization (ESRO) seems to exhibit a greater interest in scientific research whereas the European Launching Development Organization (ELDO) reflects a greater concern with the political import of its prospective activities.

Most active in giving impetus to international exchange and co-operation in space has been ICSU which established the international Committee on Space Research (COSPAR) in 1958. (For a discussion of COSPAR see pages 291–98.) Having developed a pattern for the genesis of international space science activities during the International Geophysical Year, the science unions with ICSU's encouragement and COSPAR's assistance continue to perform a vital role in this field, which assumes even greater importance in the absence of a permanent intergovernmental space agency. COSPAR is regarded by both national and international science institutions as the primary international space committee.

INTERNATIONAL COUNCIL OF SCIENTIFIC UNIONS (ICSU)

International co-operation in outer space is not as new as one would surmise from the present activities of the various international organizations interested in space affairs. Astronomy, even in early times, developed on a basis of widespread co-operation. Much of the present emphasis on outer space activity owes its origin to the rocket and satellite programs which began during the International Geophysical Year. Of significant assistance to the development of the IGY was the existence of ICSU, a ready-made organization and an outgrowth, in part, of earlier international programs in geophysics. In order to understand how ICSU performed this role it might be helpful to examine briefly some of its general characteristics.

ICSU has dual membership. One group is made up of scientific institutions, primarily national scientific academies or research councils; the other group consists of international unions devoted to a particular subject such as geophysics, astronomy, physics, or chemistry. ICSU emphasizes the unity of science by acting as a cohesive body for scientific unions and by providing a mechanism for discussion of inter-union problems. It also has become a mechanism for undertaking special programs requiring considerable inter-union co-ordination; the IGY was one such program.

In 1952, the ICSU established a special committee for the IGY known as CSAGI (from the initials of its name: Comité Spécial de l'Année Géophysique Internationale) composed of specialists from the various fields to be explored. After its first meeting, CSAGI, through ICSU, issued invitations to national science academies throughout the world to participate in the activities of the IGY, following which a succession of IGY assemblies took place during the next six years. For eighteen months—from July 1, 1957, to December 31, 1958—some fifty thousand scientists, technicians, and observers, representing sixty-six countries, engaged in an unparalleled investigation of the earth and its surroundings. Their observations and experiments were conducted in more than 4,000 stations ranging from the drifting ice in the Arctic Basin to Antarctica, and from deep marine trenches in the Pacific to satellite outposts thousands of miles above the earth.

COMMITTEE ON SPACE RESEARCH (COSPAR)

One of the outstanding achievements of the IGY was the development of a program in space research which resulted in such exciting discoveries as the Van Allen radiation belts. Even before the end of the IGY, the value of continued international co-operation in rocket and satellite research was widely recognized. This resulted in the presentation by CSAGI of a resolution to the International Council of Scientific Unions recommending that a suitable means for continuing plans and co-operation in space research be found. Thus in October, 1958, before the official end of the IGY, a special committee for space research—COSPAR—was established by ICSU for a period of one year. The primary purpose of COSPAR was to "provide the world scientific community with the means whereby it might exploit the possibilities of satellites and space probes of all kinds for scientific purposes and exchange the resulting data on a co-operative basis." COSPAR's concern, as outlined in its charter, was more with fundamental research; it was not to concern itself with the technological problems such as propulsion, construction of rockets, guidance and control.

COSPAR's objectives were to be carried out by the international community of scientists working through ICSU and its adhering national academies and international scientific unions. The original composition of COSPAR consisted of (*a*) representatives of national scientific institutions of the seven countries launching satellites or having a major program in rocket research: Australia, Canada, France, Japan, the Soviet Union, the United Kingdom, and the United States; (*b*) the representatives of the national scientific institutions of three of the countries involved in tracking or other forms of space research on an agreed system of rotation; and (*c*) representatives from nine international scientific unions.

At the second meeting of COSPAR held at the Hague during March, 1959, the member from the Soviet Union declared that he did not consider the composition of COSPAR as meeting the need of "broad international co-operation." The Soviet member proposed that the Ukraine and Byelo-Russia be added to the category of countries launching rockets and proposed six other Eastern European countries to the category of rotating membership. The proposals were not accepted, with the consequence that the Soviet Union withdrew temporarily (until 1960) from COSPAR, but never quite severed all its connections. However, at the same meeting a resolution was unanimously adopted which recommended that the composition of COSPAR should be reconsidered.

The new formula that was finally evolved provided for the following: membership to COSPAR would be open to *all* national members of ICSU that are actively engaged in space research, and to those international scientific unions that are members of ICSU and that signify their desire to participate in COSPAR. In the reorganization that took place, the executive council of COSPAR was reconstituted with a seven-man bureau composed of a president, two vice-presidents, and four additional elected members, all elected for a three-year term, and of representatives of ten international science unions.

The new COSPAR charter and by-laws were given final form

by the COSPAR executive committee during its November, 1959, meeting at Amsterdam and approved by ICSU in December, 1959, at which time COSPAR was made a special committee of ICSU, eliminating its provisional character. The Soviet Union resumed full participation by attending the third general meeting of COSPAR at Nice, France on January 8, 1960. At the Nice meeting the election of the bureau took place as did the establishment of the three permanent working groups under COSPAR which up to that time had been functioning in provisional form.

COSPAR presently consists of eighteen national scientific institutions, ten international scientific unions, an executive council, a bureau, four working groups, and a secretariat. The administrative secretary and executive secretary are directly responsible to the president of COSPAR who is Maurice Roy (member of the French Academy of Sciences).

Three requirements were stipulated for national scientific institutions to qualify for COSPAR membership. They are: (1) adherence to ICSU, (2) the desire to be represented in COSPAR, (3) present evidence of active engagement in space research on the part of the scientists in the country or territory which they represent.

The requirement "actively engaged in space research" was interpreted in a communication of November 23, 1959, from the secretary-general of ICSU to all ICSU national members:

> Active engagement in space research will be interpreted to mean scientific research in connection with the launching or tracking of rockets, satellites or space vehicles, and in the elaboration of their scientific payloads; and also research in a related scientific discipline, including instrument design.

The procedures for formal approval of national scientific institutions represented in COSPAR were established by July 26, 1960. In addition to the seven national scientific institutions which formed the original institutional membership of COSPAR, eleven more scientific institutions have joined COSPAR.

According to the COSPAR charter, each of the national

scientific institutions represented in COSPAR can send representatives to all COSPAR meetings, establish scientific channels for exchange of information, participate in obtaining data, cooperate in carrying out international space experiments, evaluate information, and make available scientific results of space research conducted as part of their participation in the work of COSPAR. The financial contribution that each is requested to make to COSPAR is recommended by the executive council, subsequently approved by COSPAR and ICSU.

While the national institutions can vote on all matters in COSPAR, the international scientific unions can vote only on matters which involve neither major items of income or expenditure by COSPAR nor considerable expenditure by the national scientific institutions. This is because the unions are not required to contribute to the financial support of COSPAR. In all other respects, generally, the rights and duties of the unions are similar to those of the national institutions. It should be mentioned that these provisions notwithstanding, COSPAR's *modus operandi* is by general consensus of all its members.

In April, 1961, a new working group structure with an expanded membership was established for COSPAR at its Florence meeting. Working group 1, formerly known as Tracking and Transmission, was renamed Tracking and Telemetering; working group 2, Scientific Experiments; working group 3, Data and Publications; working group 4, previously an *ad hoc* committee, was established as the COSPAR International Reference Atmosphere group.

The preparatory group of the International Reference Atmosphere of COSPAR has already made an extensive report which includes an integrated set of tables of upper atmosphere properties from 32 kilometers to 800 kilometers, compiled from available sources including original tabular source materials provided by the United States, the Soviet Union, and other countries. This report was distributed in August, 1961, for further comment and study. Rounding out the working groups of COSPAR is a panel on polar cap experiments which is a part of a working group on scientific experiments.

A conference held in January, 1960, at Nice, France, rep-

International Space Organizations

resented the end of COSPAR's preoccupation with questions of organization, and the conference held in Florence, Italy, in April, 1961, represented the first opportunity for COSPAR to devote undivided attention to its scientific program. At the Nice space science symposium, 98 scientific papers were presented from nine countries with approximately 250 scientists in attendance from eighteen countries. All of these papers were published subsequently in a single volume entitled *Space Research*. The Florence meeting discussed radio and optical tracking, magnetic observation by rockets and satellites, telemetry and data recovery, special events, recent results from instrumented satellites and space craft, international reference atmosphere, and scientific research by means of small sounding rockets. Together with the reports made by the national scientific institutions and by the international scientific unions, a total of 109 scientific papers were presented with a total participation of 290 people. These reports and papers are being published. In May, 1962, the fifth COSPAR plenary session was held in Washington, D.C., where a series of internationally co-ordinated space experiments such as rocket soundings were proposed. A consultative group on potentially harmful effects of space experiments was established, as were various procedures to further the standardization and unification of symbols, units, and nomenclature in space science.

COSPAR, as one of the offspring of the IGY, has not only continued many of the IGY space activities but has extended the IGY pattern of space co-operation. COSPAR, however, is not an agency which initiates independent investigations of space science, nor is it a governing body to approve whatever research might be taking place. Its major objective is to maintain the spirit of scientific co-operation in fields closely related to space research, and to provide a meeting place for exchange of scientific initiative from all directions.

COSPAR, by its very existence, has offered itself as a forum for the discussion of experimental ideas and the presentation of various space projects. At a COSPAR conference the United States Academy of Sciences proposed on behalf of NASA that COSPAR members place in a United States-launched satellite individual experiments or complete payloads of mutual interest prepared by scientists of other nations. NASA has reaffirmed its readiness to make available launching vehicles, spacecraft, technical guidance, and laboratory support for useful experiments or payloads developed by foreign scientists. All COSPAR members were invited by NASA and by the United States Weather Bureau to participate in meteorological research connected with one of the Tiros satellites.

In addition to these concrete offers, the Space Science Board of the United States National Academy of Sciences continues to send to COSPAR dispatches of preliminary technical information on launchings of rockets and satellites, regular transmittals of orbital and satellite elements through Spacewarn (an international scientific network operated under the auspices of COSPAR), and deposits reports on scientific results in the IGY World Data Centers.

Indicative of the interest in COSPAR's views is the fact that COSPAR has been invited to attend almost every important international science gathering which dealt with matters related to space research. COSPAR has been represented in meetings, conferences, and symposia on space research which have dealt with rockets and astronautics (such as those established by the International Astronautical Federation), space medicine, telescopes in space, space radio, and geophysics.

COSPAR also consulted with the organizers of the European regional space conference which took place in London on April 29, 1960, with representative space scientists from ten European countries who discussed co-operation on space science. The specific question of COSPAR's relationship to this projected organization was discussed. Van de Hulst, former president of COSPAR, explained that COSPAR's charter provided that "as a non-political organization COSPAR shall not, as a matter of policy, recommend any specific assistance of any one nation by another. It will, however, welcome information concerning such arrangements and provide a convenient assembly in which such arrangements may formally be proposed and discussed." He added that if the ten Western European nations worked out any continuing plan for multilateral cooperation in space research, COSPAR would gladly extend the same assistance for

International Space Organizations 249

the exchange of scientific information to this body which it already provides to all other countries conducting scientific experiments with sounding rockets and space vehicles.

Among the activities of COSPAR are its sponsorship of the international rocket intervals. These rocket intervals consist of rocket firings co-ordinated on a world-wide basis to observe atmospheric properties. The first rocket interval organized by COSPAR with the national space research committees took place in November, 1959, and the second in September, 1960. Beginning in 1961 international rocket intervals have been held in January, April, July, and October of each year.

COSPAR has been called upon to assist two international co-operative programs, the World Magnetic Survey (WMS) and the International Year of the Quiet Sun (IQSY), by providing technical recommendations for rocket and satellite experiments in support of the objectives of these programs. The WMS program was a project of the IGY deferred until after the time of maximum solar activity and continuing until 1965, when new charts of the geomagnetic field will be compiled. A committee of the International Association of Geomagnetism and Aeronomy is responsible for technical recommendations for WMS while international co-ordination is under the aegis of the ICSU's International Geophysical Committee (CIG), the successor to CSAGI. The CIG is also responsible for developing and co-ordinating the program for IQSY, planned to take place during a period of minimum solar activity in 1964–65 with the support of the International Union of Pure and Applied Physics, the International Union of Geodesy and Geophysics, the International Astronomical Union, and the International Scientific Radio Union. In May, 1962, COSPAR urged broad participation in a Sunspot Minimum Ionospheric Rocket Sounding Program during the IQSY. For the first time, synoptic properties of the "quiet" ionosphere would be obtained, providing a better understanding of the physical processes taking place in this region.

Among its varied activities COSPAR issues a quarterly bulletin for authoritative, timely international communication on space matters. COSPAR also transmits information on orbital elements, technical details of planned launchings, and telemetry codes of some space experiments so that scientists in many countries can make observations and participate in tracking. COSPAR has examined the existing World Data Center structure established for data exchange under the auspices of the IGY and continued under CIG, and discussed possible improvements in the exchange of rocket and satellite data. COSPAR has studied such topics as space needs for radio frequency allocations, the properties of the upper atmosphere, and the remarkable geophysical events of July, 1959, and November, 1960, associated with solar activity and events of the upper atmosphere.

Among the intangible activities of COSPAR are the personal exchanges among scientists and the comparison of scientific papers presented to conferences and for publication. COSPAR encourages informal discussion on space research not only among individuals but among national and international institutions and countries; but it refrains from specific recommendations in order to leave decisions completely to the countries or institutions involved.

COMMITTEE ON CONTAMINATION BY EXTRATERRESTRIAL EXPLORATION (CETEX)

Alert to the possibilities of contamination of outer space by vehicles launched from earth, the scientific community sought to deal with this problem, and in 1958 the International Council of Scientific Unions decided to establish a committee on this subject. Seven scientists were appointed to CETEX, including representatives from the United States and the Soviet Union.

In an attempt to establish the dimensions of the problem, CETEX sponsored a conference of scientists in May, 1959. The result of this conference was an international code governing the control of contamination of celestial bodies. Specific recommendations of this conference included:

> Deferral of nuclear explosions on the moon which could make it impossible for future space explorers to study such questions as the origin of lunar rocks through analysis of their radioactivity.
> No soft landing on the moon should be attempted until a thor-

International Space Organizations

ough study has been made of its atmosphere with low flying satellites. This is important because the tons of fuel used in braking a rocket to a gentle landing would make it impossible to tell the true nature of the lunar air which is purported to be relatively thin.

No big TNT blasts should be set off inasmuch as it would contaminate the lunar atmosphere for years.

Hard landings with rockets should be confined to limited areas to avoid extensive contamination of the moon with large molecules produced by life on earth.

CETEX pointed out the possibility that some organisms found on earth might grow and reproduce on other planets. This would be most likely on Mars, where the necessary ingredients of water, carbon dioxide, nitrogen, and light are available for photosynthesis. While the moon provides a less fertile environment for the multiplication of cells because of an absence of water and no significant atmosphere, there still exists the possibility of contaminating the moon with terrestrial organisms which might interfere with any "pre-life" molecules that might be present. One intriguing question which could shed light on the origin of life is how these complex "pre-life" molecules come into existence, and how they reproduce. Molecules from earth could seriously interfere with answering this question. Consequently, dead bacteria would be almost as harmful as live ones. This underscores the immense difficulty, if not impossibility, of sterilizing the rockets adequately before launching. The problems posed by adequate decontamination are quite complex and have posed difficulties which are the subject of present study. Nevertheless the importance of effective decontamination has been widely acknowledged and is manifested by the absorption of CETEX by COSPAR's working group on Scientific Experiments.

THE SCIENTIFIC UNIONS

The principal non-governmental international agencies which have demonstrated some interest in space research are the international scientific unions, especially in fields which benefit from experiments with sounding rockets, satellites, or space probes.

Space activities promoted by these organizations are, thus far, extensions of their fields of interest and not a new scientific discipline. This has brought together scientists of different disciplines who are interested in studying outer space. It has also required co-ordination between these disciplines both in basic and applied research.

The scientific unions particularly interested in space research are: the International Astronomical Union (IAU), the International Union of Geodesy and Geophysics (IUGG), the International Union of Pure and Applied Chemistry, (IUPAC), the International Scientific Radio Union (URSI), the International Union of Pure and Applied Physics (IUPAP), the International Union of Biological Sciences, the International Union of Theoretical and Applied Mechanics, the International Union of Physiological Sciences, and the International Union of Biochemistry.

Some idea of the interests of these unions can be gained from a brief summary of the interests of five international science unions. The International Astronomical Union's interests in space consist primarily of radio frequency needs for space science studies and protection of frequencies for radio astronomy (done in co-operation with URSI and COSPAR); ultraviolet and X-ray studies of the sun, stars, and the galaxies; celestial mechanics; and physics of the moon, planets, meteorites, and comets. Many commissions of the IAU have recently shown a marked increase in activity, largely because of the impact of space research. This renewed interest in space research was demonstrated in the August, 1961, IAU meeting (Berkeley, Calif.), especially in the reports of four commissions: celestial mechanics, physical study of comets, physical study of planets, and light of the night sky. In addition, a new commission on observations from outside the earth's atmosphere was organized in 1959. It held its first full-scale session in 1961. One IAU commission has proposed hitting a comet with a space probe to determine the constituent elements of the comet. The IAU has continued to exercise its traditional function of standardizing terminology. This area has been affected by the recent increase in space research, especially in celestial mechanics, which many young newcomers have entered with varying sets of terms and practices.

The International Radio Science Union (URSI) has an active space radio committee which monitors techniques of communication and scientific aspects of radio propagation and advises URSI on its role. URSI, in turn, advises COSPAR on radio research and co-operates with COSPAR in requesting or allocating radio frequencies for space research. An URSI symposium in which COSPAR joined was held in Paris in September, 1961, and dealt with scientific aspects of space radio communication problems. Another symposium, held in 1962, discussed maximizing information transfer from space vehicles.

The International Union of Geodesy and Geophysics (IUGG) consists of seven specialized associations. Three of these have been particularly active in space research: the International Association of Geomagnetism and Aeronomy (IAGA), the International Association of Geodesy (IAG), and the International Association of Meteorology and Atmospheric Physics (IAMAP). After the meeting of the IUGG General Assembly in Helsinki, Finland, in August, 1960, IUGG established a committee on space research to provide effective co-operation with COSPAR on all subjects of mutual interest. An instance of this co-operation is the symposium on the meteorological uses of rockets and satellites held in Washington, D.C., in April, 1962, under the joint sponsorship of WMO, IUGG, and COSPAR.

The International Union of Theoretical and Applied Mechanics (IUTAM) is interested mainly in symposia on space and atmospheric matters that have direct bearing on mechanics. The IUTAM and the IUGG held a joint symposium at Marseilles, France, on September 19, 1961, on fundamental problems in turbulence and their relationship to geophysics. The principal aim of this symposium was to study turbulence in a gravitational field characterized by vertical gradients of density and of temperature.

The International Union of Pure and Applied Physics (IUPAP), whose interest in outer space predated the IGY, was especially concerned with cosmic rays and magnetometry, as well as developments of the rocket and satellite program during the IGY. IUPAP's continuing interest in cosmic space is pursued by its Commission on Cosmic Rays.

To provide an opportunity for scientists from many countries to discuss specific international experiments in the fields of cosmic rays, energetic solar particles and geomagnetism (which depend on world-wide coverage or synoptic observations), several informal working group meetings took place at Kyoto, Japan, in September, 1961, as adjuncts to the international conference on cosmic rays and earth storms held under the auspices of IUGG, IUPAP, URSI, and IAU.

INTERNATIONAL ASTRONAUTICAL FEDERATION (IAF)

The idea for an international federation of astronautical societies was first broached in 1949 at the summer meeting of the then newly formed astronautical society (Gesellschaft für Weltraumforschung) of West Germany which recommended convening "an international meeting of all societies for rocket development, interplanetary travel and space research, to foster friendly relations and a successful exchange of knowledge and to explore the possibilities of forming an international association for astronautics." In 1951 at London, the International Aeronautical Federation was formally inaugurated as an association of interplanetary and rocket societies, composed of amateurs, enthusiastic laymen, and some professionals interested in interplanetary matters.

A large segment of IAF membership would like to go beyond conferences and promotional activities and establish an international research institute for astronautics. However, the more predominant view has been that the state of international relations did not lend itself to the establishment of such a research institute; hence, this has been continually postponed. Renewed interest in this project has resulted in the recent establishment of the International Academy of Astronautics with marked interest in a possible international lunar laboratory. The Academy was established under IAF auspices on August 15, 1960. Theodore

International Space Organizations

von Karman, aerodynamicist, was appointed director until the first regular meeting of the Academy could take place. The Academy's first symposium was conducted in France during June, 1961, and discussed space flight and re-entry trajectories.

Since its inception, IAF has collaborated informally with several international bodies to facilitate co-operation under law, and has discussed various legal topics in many of its symposia. The establishment of the International Institute of Space Law by the IAF represented the culmination of IAF's interest and activity in this area. The purpose of this institute has been to foster the social scientific aspects of astronautics by holding meetings and colloquia,to make studies and reports and the publication thereof, and to make awards.

UNITED NATIONS

The original interest of the United Nations in outer space was first manifested in disarmament proposals. On January 14, 1957, a disarmament plan was formally presented to the United Nations by United States Ambassador Henry Cabot Lodge. One of the five points stipulated that there should be U.N. regulation to assure that outer space is used only for peaceful purposes. Registration of launching vehicles seemed to be a first step toward more widespread international regulation and control. International inspection was proposed in conjunction with space regulation embracing satellites, intercontinental missiles, space platforms, and long-range unmanned weapons. These matters were considered by the U.N. disarmament committee and a resolution subsequently adopted by the General Assembly on November 14, 1957, over the objection of the Soviet bloc, called for a study of the United States proposal.

A Soviet counterproposal linked the use of outer space to United States use of foreign military bases, and within this general approach asked the thirteenth General Assembly in 1958 to establish a U.N. agency for international co-operation in the study of outer cosmic space. The United States also proposed an outer space topic for the agenda which was subsequently combined with the Soviet proposal.

After contemplating the Soviet resolution, the committee considered a United States resolution which had the formal endorsement of nineteen other nations: Australia, Belgium, Bolivia, Canada, Denmark, France, Guatemala, Ireland, Italy, Japan, Nepal, the Netherlands, New Zealand, Sweden, Turkey, the Union of South Africa, the United Kingdom, Uruguay, and Venezuela. This twenty-power resolution called for the establishment by the General Assembly of an *ad hoc* committee on the peaceful uses of outer space to report to the next session of the General Assembly on the activities and resources of the U.N. and other international bodies, international outer space programs that could be taken under U.N. auspices, and future U.N. space organization arrangements, as well as legal space problems.

In a drastically revised version of its resolution of November 18, the Soviet Union eliminated mention of a U.N. agency and of military bases. It, too, recommended the establishment of a U.N. committee for the study of outer space with an eleven-nation preparatory group: the Soviet Union, Czechoslovakia, Poland, Rumania, the United States, the United Kingdom, France, Argentina, India, the United Arab Republic, and Sweden. This committee was to continue the spirit of co-operation engendered by the IGY, to organize an exchange of information and to co-ordinate national space research programs.

The twenty nations reacted favorably to the proposal, but objected to the Russian proposed composition of the committee, countering with their own recommendation for eighteen members: all of the members proposed by the Soviet Union, except Rumania plus Australia, Canada, Iran, Mexico, Belgium, Italy, Brazil, and Japan.

Since the issue of committee composition could not be satisfactorily resolved, the Soviet Union withdrew its draft resolutions stating that unanimity in this matter was essential. By a vote of 54 to 9 (Soviet bloc) with 18 abstentions, the twenty-power draft as revised was subsequently adopted. Immediately, however, the Soviet Union, Poland, and Czechoslovakia announced that they would not co-operate in the *ad hoc* committee's work. On December 13, 1958, the General Assembly by

International Space Organizations

resolution 1348 (XIII) adopted the proposal to establish an *ad hoc* committee on the peaceful uses of outer space to report to the following session of the General Assembly.

On May 6, 1959, after many unsuccessful attempts to secure Russian participation, the U.N. *ad hoc* committee met. In view of the Soviet bloc's absence, both the United Arab Republic and India also declined to take part as members of the U.N. committee. The abridged committee nonetheless considered some problems of outer space, but did not go so far as to advance any solutions.

On June 25, 1959, the full *ad hoc* committee submitted its final report. In the section of this report dealing with technical and scientific activities, the stress was on co-operative efforts in space research, in which it was felt that COSPAR would have a particularly important role to play. It was generally agreed that when the research stage was passed, functional intergovernmental arrangements concluded through specialized agencies such as WMO, ITU, ICAO, UNESCO, and others would probably be desirable. In recognition of the continuing interest of specialized agencies in outer space activities, the *ad hoc* committee suggested that the General Assembly might ask these groups to include all information on their space activities in their annual reports to the United Nations.

For further discussion of this report see page 227. This report was subsequently considered by the General Assembly which on December 17, 1959, established a twenty-four member Committee on the Peaceful Uses of Outer Space to serve during 1960 and 1961. Hampered by Soviet failure to participate, the space committee was inactive during the course of these two years.

On December 11, 1961, the General Assembly voted unanimously to continue the Space Committee (adding Chad, Ghana, Mongolia, and Sierra Leone) and requested the enlarged committee to meet not later than March 31, 1962, to consider weather and communications, legal space problems, contacts with governmental and non-governmental organizations, exchange of information and promotion of international space cooperation. The resolution states that "(*a*) International law, including the United Nations Charter, applies to outer space and celestial bodies; and (*b*) Outer space and celestial bodies are free for exploration and use by all states in conformity with international law, and are not subject to national appropriation." Nations were asked to furnish the United Nations with information on registration of launchings for a public registry.

Member states and the WMO were requested to study measures for the advancement of atmospheric science and technology. The WMO, in consultation with other organizations including ICSU, was asked to make organizational and financial recommendations to achieve these ends through the Space Committee.

The General Assembly also invited the U.N. Technical Assistance Program and Special Fund, in consultation with ITU, to consider requests from member states for surveys of their communication needs and facilities. ITU was asked to consult with other states and organizations, including COSPAR, and to submit a report through the Space Committee.

SPECIALIZED AGENCIES

In mid-November 1962, the NASA and U.S. Weather Bureau convened an International Meteorological Satellite Workshop in Washington, D.C., which was attended by twenty-seven nations. The U.S.S.R., Poland, and Czechoslovakia were invited and had indicated acceptance, but subsequently declined to attend.

This negative turn in the development of space co-operation was followed by a significant and positive advance with the submission by the United States and the U.S.S.R. of data to the U.N. for its public registry on objects launched into orbit and beyond. In conformance with the U.N. resolution of December, 1961, the United States and the U.S.S.R. have submitted periodic reports on their respective launchings for the U.N. registry through April, 1962.

Another favorable occurrence during this period was the exchange of letters between President Kennedy and Premier Khrushchev during February and March, 1962, calling for wider international space co-operation. In his congratulatory letter on

International Space Organizations

John Glenn's flight, Khrushchev recommended a pooling of space efforts for the benefit of mankind. In his reply, Kennedy specified five areas of potential co-operation: weather forecasting, magnetic field studies, space medicine, communications exchange, and radio tracking stations, and further suggested joint efforts in manned and unmanned exploration of the moon and other celestial bodies. Khrushchev, alluding to the huge costs of space exploration, offered three other areas: search for and rescue of disabled spacecraft, a study of legal space problems, and an agreement to restrict experiments that interfere with peaceful exploration.

Following this exchange, the U.N. Committee on the Peaceful Uses of Outer Space met during the last week of March, 1962, and discussed the proposals raised by Kennedy and Khrushchev as well as other ways to promote space co-operation. While no formal resolution was passed, the committee did approve (without a vote) the report of its chairman which summarized the views expressed by the members during the session of the need for space co-operation, expressed satisfaction at the Kennedy-Khrushchev exchange, and the desirability for co-ordinating the space work of the relevant governmental and non-governmental agencies.

As a follow-up to the March session, the technical and legal sub-committees of the U.N. Space Committee convened on May 28, 1962, in Geneva to discuss specific ways of furthering space co-operation. The legal subcommittee debated such items as a legal space code and safe return of astronauts. The technical subcommittee constituted three working groups on type of information for furthering space co-operation including the functions of world centers to collect data, on concrete international programs of scientific research and on the desirability of setting up an international rocket launching facility near the geomagnetic equator. The reports of the subcommittee have been forwarded to the full committee, which met at the U.N. in September, 1962.

Specialized agencies have evinced varying degrees of interest in outer space according to their objectives and activities. Thus far each specialized agency has developed this interest in a relatively informal manner, but has established contact for purposes of co-ordination both with the U.N. through its Economic and Social Council and with COSPAR. Currently, there are six specialized agencies which have a predominant interest in space affairs: the International Telecommunications Union (ITU), the World Meteorological Organization (WMO), the United Nations Educational, Scientific, and Cultural Organization (UNESCO), the International Civil Aviation Organization (ICAO), the World Health Organization (WHO), and the International Atomic Energy Agency (IAEA). For discussion of ITU and WMO space-oriented activities, see pages 281–83.

As its name implies, UNESCO is concerned with educational, scientific and cultural activities having international ramifications. UNESCO has a department of natural sciences but it lacks laboratories and scientific equipment and thus does not conduct any research on its own. UNESCO's *modus operandi,* accordingly, is to advance world scientific efforts through the auspices of existing international scientific organizations, both governmental and non-governmental, such as ICSU, COSPAR, UN, and IAF, and the International Council of Medical Sciences. While UNESCO's major function in "cosmic" matters as manifested thus far is to support the activities of other agencies in space affairs, it has served to stimulate international co-operation on a regional and sometimes world-wide basis.

As early as 1951, UNESCO assisted ICSU in establishing the IGY. Interested in co-operating in scientific projects of mutual interest and feeling that the IGY scientific projects conformed to the general objectives of UNESCO, it provided an annual subvention to ICSU in 1951. By 1958, UNESCO, in addition to national committees for the IGY, had contributed over $89,000 to help defray the expenses of CSAGI.

UNESCO has aided COSPAR financially in carrying out its 1959 meeting at Nice, France and has consistently sent observers to the various meetings of COSPAR, both as an indication of its interest and as a means for insuring co-ordination between the two organizations. During the general conference of UNESCO held in November–December, 1958, the following resolution was passed:

International Space Organizations 261

The director-general is authorized . . . to study scientific problems, the solution of which may help to improve the living conditions of mankind, to stimulate research on these problems, and to promote when appropriate the adoption of international or regional measures for the development of such research, particularly, in the following fields: . . . (i) exploration of extraterrestrial space. . . .

Thus far UNESCO's major interest in space affairs vis-à-vis other organizations has been mainly in the area of information and publication in order to stimulate interest in research in outer space.

The interest of the International Civil Aviation Organization (ICAO) has centered to date around legal aspects of space problems. In February, 1959, the ICAO council brought to the attention of its assembly which met on June 16, 1959, the suggestion that ICAO study the legal status of outer space and the regulation of spacecraft, particularly with reference to the traffic of civil aircraft in air space. However, it was pointed out that the question of the legal aspects of outer space was one of the matters under consideration by the U.N. and that ICAO should accordingly take into account the need for co-ordination with the deliberations of the U.N.

While the legal subcommittee of the U.N. *ad hoc* committee on outer space did not arrive at solutions to legal space problems, it did describe what these problems were and suggested how they might be solved. However, it would appear that neither the U.N. nor its agencies, including the ICAO, has taken any specific action in this field since 1959.

While the World Health Organization (WHO) has not conducted any specific work related to outer space nor does it contemplate any in the near future, the organization can be very useful in the stimulation of research, the publication of medical findings, and the holding of symposia and seminars pertinent to medical or health problems associated with space exploration and travel.

Similarly, the International Atomic Energy Agency (IAEA) does not contemplate any concrete effort in the field of outer space in the very near future. However, IAEA has expressed an interest in the nuclear technology of outer space and might be in a position to advise on aspects of its health and safety. The United States Atomic Energy Commission established a panel in November, 1958, to examine the effects of space radiation on health, especially radiation hazards from the use of atomic-powered vehicles in space. Interdisciplinary in its approach, the board includes specialists in biology, genetics, fallout, law, international affairs, licensing, regulation, radiological health and safety. The assigned task of this panel is to study the possible effects of nuclear space devices on the health of the people of the world and to recommend standards for safe practices.

SPACE AGENCIES IN EUROPE

There are three European space organizations in the process of formation: European Space Research Organization, European Launcher Development Organization, and Euro-Space.

Euro-Space is a convention of European industries specializing in space research and development. In May, 1961, concerns engaged in rocket research in France, the United Kingdom, Italy, the German Federal Republic, Belgium, Holland, and Sweden set up a private committee for European space research at the instigation of the British Hawker-Siddebey concern and the French Societé pour l'Etude et la Réalisation d'Engins Balistiques (SEREB). The inaugural meeting was attended by thirty-five scientists from thirty private concerns in the aircraft, electronics, and chemical fields. Working groups will deal with long-range communications for civilian purposes, navigation satellites for civilian purposes, and anti-satellite systems for military purposes. The first general meeting of Euro-Space was held in Paris in September, 1961, when the first reports on the proposed European space research program were submitted.

The entire momentum for European interest in consolidated space efforts stemmed from a rejection of the policy of "going it alone," not only by the lesser developed countries of Europe but also by such countries as the United Kingdom and France. Anticipating that future research tasks would far exceed their individual financial resources, the Western European nations felt that co-operation was essential. With the significant work already accomplished by Britain and by France in this field, the Western European nations also felt that a consolidation of effort in space research and rocket launchings would lead to more effective accomplishment.

During the early part of 1960, interest in determining what role Europe could perform in space research was the source of discussion at an informal gathering of European scientists convened under the initiative of Professor Edvardo Amaldi of Italy and Professor Pierre Auger of France. The Royal Society of England in April, 1960, invited scientists of ten other European nations to meet in London and discuss the possibilities of European co-operation in space research. A series of meetings followed which saw the establishment in June, 1960, by the ten Western European states, of the "Groupe d'Etudes Européennes pour les Recherches Spatiales" (GEERS) which made an official proposal to national governments to form a space research organization. In November, 1960, Spain joined the original ten European nations (Belgium, Denmark, France, Italy, the Netherlands, Norway, Sweden, Switzerland, the United Kingdom, and the German Federal Republic) which met in Meyrin, Switzerland, for four days. Envisioned as a sort of counterpart to Western Europe's nuclear research center (CERN), the projected European space agency would concentrate on satellites and research rather than on launching rockets, and would undertake research which would be too costly, too complicated, or too widespread to be conducted properly by any one nation alone. Like CERN this organization would make public the results of its research for the benefit of scientists everywhere.

As a result of the November session in Switzerland, the eleven-nation delegation approved the establishment of the European Preparatory Commission for Space Research (COPERS) with the responsibility for organizing the space agency within one year. The commission's task was to work out the details concerning the nature of the organization, to prepare its scientific and technical program, and to choose the site of the agency's permanent headquarters and other facilities. Like the European space research agency just discussed, this projected organization is designed to supplement rather than duplicate United

States or Soviet space efforts, although of necessity, some activities will be paralleled. While the primary concern of the research agency is with space research, that of the launching organization is with rockets as well as with industrial and commercial application of satellites.

At a meeting held in Strasbourg, France, in January, 1961, under the auspices of the Council of Europe and at the invitation of Britain and France, twelve nations convened to deliberate on the feasibility of establishing an European space launching program. (These included the same eleven nations as in the projected European Space Research Organization, plus Austria.) The final committee report published by this conference recommended launching three satellites: a stabilized satellite weighing 1,000 to 2,000 pounds to be put into a low orbit, mainly for astronomical observations; a smaller satellite weighing roughly 250 pounds to be placed in an eccentric orbit, for investigation of the earth's gravitational and magnetic fields, cosmic radiation, and exosphere at distances up to three times the earth's radii; and a 100-pound satellite to orbit in a highly eccentric trajectory at a distance of up to twenty-five times the radius of the earth to investigate the solar outer corona and possibly be used for a telecommunications relay station. The report also considered future possibilities of rocket propulsion, including the use of nuclear and solar energy as well as ion beam propulsion, and the technical, scientific, and economic benefits which Europe could derive from space research.

The Strasbourg conference further recommended that a three-stage booster be built with the British Blue Streak as the first stage, a developed French Veronique as the second stage, and a third stage to be built by several firms, probably in continental Europe. The other European nations would develop and manufacture equipment for launching and payloads, as well as for tracking and data acquisition, and any additional base facilities that would be desired. It was thought that while test firings of the separate stages could begin in about two years' time, the first fully tested booster would not be able to put a satellite in orbit until about 1965.

While both European space organizations are still in the

early stages of development, they do underline Western Europe's interest in broad co-operation in space affairs. Although much background work has taken place during their development, there remain significant practical difficulties which require resolution before these organizations can function effectively.

LATIN AMERICA

Interest in space research has been manifested in several Latin American countries as well as in inter-American discussions on space research, notably the symposium sponsored by the Comision Nacional de Investigaciones Espaciales conducted in Buenos Aires in December, 1960. Scientific papers presented at this symposium included such subjects as the use made of balloons, rockets, and satellites for meteorology and communications, and scientific studies concerned with terrestrial and cosmic phenomena. There was an especially stimulating discussion on the unusual solar eruptions which had occurred only a few weeks before the symposium. The fact that significant data was presented at such an early date stands out as an auspicious example of the timely and successful sharing of scientific opinion and data with an audience of scientists from different disciplines and geographical areas.

One concrete result of this symposium was the establishment of a non-governmental provisional Inter-American Committee for Space Research (ICSR) with the following purposes:

1. To promote the formation of local committees composed of natural and social scientists and engineers connected with space science and technology who would be encouraged to use the facilities of ICSR.
2. ICSR will send all available information to the local committees and encourage them to establish National Commissions or obtain government support for increased space research.
3. Frequent local meetings would be held.
4. Periodic bilingual publications with abstracts of papers will be published.
5. Study of space science and programs in research centers and universities will be promoted.

The permanent committee has not yet worked out a concrete program, but there has been an exchange of correspondence among members with the encouraging consequence of strengthening space research in Latin American universities, notably in Argentina.

INTERNATIONAL PROGRAMS OF NASA

ARNOLD W. FRUTKIN

International co-operation in the conduct of the nation's space research and exploration programs, vigorously endorsed in the President's State of the Union address, is a specific objective of the National Aeronautics and Space Act of 1958. There, the Congress stated that "the aeronautical and space activities of the United States shall be conducted so as to contribute materially to . . . co-operation by the United States with other nations and groups of nations in work done pursuant to this act and in the peaceful applications of the results thereof. . . ."

One does not have to seek far for the considerations which led Congress in 1958 to impose a mandate for co-operation upon the new civilian space agency. Clearly, some impetus toward a policy of co-operation was derived from the useful precedents of the International Geophysical Year, then much in the minds of those framing the legislation to govern the country's space program. There was, and is, promise in stimulating intellectual commerce among the nations. It was, and is, important that man's first steps into space be taken openly and in concert rather than in the dangerous context of an extension of the cold war. Beyond this, the scientific benefits to be derived from exchanges of information, from joint attacks upon the problems of space research, and from the pooling of personnel, material and financial resources—all of these were clearly recognized and require no explication here. It is sufficient to say that no one, then or now, seriously questions the intellectual and technological benefits which may reasonably be expected through an international response to the infinite challenge of space.

With the nation's dedication to international co-operation in its space activities clear, the task of implementation began.

With an eye to difficulties encountered in other programs of international character, the National Aeronautics and Space Administration formulated guidelines designed to assure substantive accomplishment in international co-operation:

So that promise will not exceed fulfillment, NASA seeks to define proposed international projects through informal technical discussion prior to reaching formal agreement with co-operating nations. This procedure is useful in achieving the purposes of a second consideration, namely, that projects have valid scientific content. (This in turn facilitates support consistent with NASA's own programs.)

The multiplicity of individual and agency interests in space research abroad (much as in this country) and the importance of assuring adequate, sustained support for very costly space research programs dictates a third guideline: that international projects be sponsored or supported centrally by the co-operating governments. The role of American dollars in NASA's international programs also had to be faced. It was decided early that both political and scientific objectives would be served better if co-operative programs were carried out without an exchange of funds between nations. Instead, each nation provides funds for that portion of a given co-operative program which represents its own commitment of staff or material. The contributions of each need not, however, be equivalent, but it is obvious that each gains by reason of the other's effort.

Proposals should be specific and reflect mutual interests and capabilities. In the optimum case, they would represent experiments or other projects which NASA itself would wish to carry out if they were not to be done jointly. Scientific results of co-operative enterprises should be generally available to the scientific community, consistent with the interests of the prime experimenters in publishing the results of their own work.

In accordance with the foregoing philosophy, NASA has embarked upon a wide range of international activities which reflect the efforts of many offices throughout the agency.

International Programs of NASA

ROCKETS AND SATELLITES

In March, 1959, the United States National Academy of Sciences' delegate to the Committee on Space Research (COSPAR) offered on behalf of NASA to place in orbit individual experiments or complete satellite payloads, of mutual interest, prepared by scientists of other nations. Since then, NASA has affirmed and reaffirmed its readiness to make available launching vehicles, spacecraft, technical guidance, and laboratory support for useful experiments or payloads developed by foreign scientists. The launching vehicle provided (without cost) may be Scout, Thor-Delta, or another, as appropriate.

The first satellites in this program are already being prepared by the United Kingdom and Canada for launching by NASA sometime in 1962. The first of three British satellites (successfully launched on April 26, 1962) carried environmental experiments (cosmic rays, ion mass spectrum, electron density and temperature, and solar radiation) while the Canadian satellite will sound the ionosphere from above. Preliminary discussions are under way looking toward the launching by NASA of satellites prepared by still other nations.

The long leadtime required for the preparation of co-operative satellite experiments allows NASA sufficient margin to initiate arrangements for the necessary launching vehicles, after agreements have been reached on the experiments. Moreover, as payload capabilities increase materially in the next two years and payload design becomes more standardized, space for individual experiments should become available in the large orbiting astronomical or geophysical satellite observatories which are now being designed.

The preparation of total satellite packages by foreign groups may be assisted initially by NASA through the provision of structural, power, or telemetry elements, but it is expected that such groups will thereafter assume responsibility for these elements. The performance by foreign scientists of a maximum of the preparatory work on their own experiments in all aspects of this program will accelerate the development of their technical capabilities, increase their potential for substantial contributions to the art, and minimize the service load upon NASA centers.

It is important to note that the very closest working relationships are required for such projects because of the rigid design and test requirements necessary to assure compatibility in mission, structure, and electronics between satellite and vehicle systems as well as between the various components of the satellite itself. Thus, in the joint satellite programs now in preparation with the United Kingdom, working parties on each side are in constant communication and meet regularly as a joint working group to resolve, by mutual agreement, the numerous technical problems inherent in preparing actual space research systems. Precisely the same procedure is in effect in the similar program now in preparation jointly with Canada.

It will often be found desirable to test proposed satellite experiments by earlier flights in sounding rockets. Indeed, scientific work with various sounding rockets is useful not only in itself but as a means for familiarizing technicians with many of the problems encountered in designing and adapting instrumentation for the conditions of space flight.

NASA encourages and assists in the development of foreign sounding rocket programs and co-operates in the activities of foreign rocket teams where their objectives appear to contribute to the over-all goals of space research, assuming again that the co-operating groups are willing to commit resources of their own to such work. In particular, NASA welcomes and encourages sounding rocket programs of synoptic value or of special geographic significance. The upper air experiments utilizing grenades and chemical reagents are especially suitable for the initial phases of foreign programs since they do not require the most complex optical or radio ground instrumentation.

In the continuing program of the Italian Space Committee, for example, a series of launchings has been undertaken to create sodium vapor clouds for the measurement of winds and temperatures in the high atmosphere. Successful tests, some synchronized with launchings at Wallops Island, have already been conducted in Sardinia and further synoptic launchings are planned. In this program, the Italian Space Committee arranged for the necessary rockets, established the launching site and con-

International Programs of NASA 271

ducted the launchings, provided optical instrumentation to retrieve the data, and is reducing and analyzing the data. NASA sponsored the Italian purchase of rockets in the United States, provided a basic launcher, and contributed the payload. Technical advice was also afforded.

In other joint programs now under way, (with Argentina, Australia, Canada, France, Italy, Japan, Norway, Pakistan, and Sweden) NASA contributes the rocket vehicles while the co-operating nations provide the payloads. The mode of co-operation is, naturally, very flexible. In any event, it is perhaps to be expected that sounding rocket programs may make up a significant portion of international space activities, especially in countries in which available funds may limit more ambitious satellite projects. Such activities are of great potential scientific value, since the restriction of sounding rockets to vertical profiles means that a multiplicity of efforts is required if comprehensive results are to be achieved. Certainly, small sounding rocket efforts should contribute economically to substantive advances in know-how in the fields of instrumentation and experimentation for space research.

GROUND-BASED SUPPORT

The most constructive contribution to space research within the present capabilities of scientists abroad may lie for some time in supporting research conducted from the ground. A program of this type was arranged in connection with the utilization of Echo I and, with the co-operation of French and British facilities, resulted in the first transatlantic communication by means of artificial satellites. Four nations are building major ground terminals for co-operative testing of early experimental communications satellites.

A similar, more extensive program was organized jointly with the United States Weather Bureau in connection with Tiros II; foreign weather services were invited to conduct meteorological observations synchronized with the passes of the satellite and to analyze the data from both sources. Instrumentation difficulties restricted the program but a valuable organizational precedent was established and will be implemented again with the launching of Tiros III. Participation in the meteorological satellite program includes twenty-nine nations: Argentina, Australia, Belgium, Brazil, British East Africa, Chad, China, Colombia, Costa Rica, Czechoslovakia, El Salvador, Federation of Rhodesia and Nyasaland, France, Hong Kong, Iceland, India, Ireland, Japan, Mauritius, the Netherlands, Netherlands New Guinea, New Zealand, Portugal, South Africa, Sudan, Switzerland, Trinidad, the United Arab Republic, and the United Kingdom. As part of the Tiros program, meteorologists from more than twenty-five of the world's weather services accepted invitations to attend an International Meteorological Satellite Workshop held in Washington, D.C., in November, 1961. Its purpose was to provide participants with information and practice in the use of satellite data for weather analysis, looking toward the time when United States weather satellites will continuously send out information available to any country that wants to receive and use it. Participants were: British East Africa, Canada, Chad, China, Czechoslovakia, Denmark, El Salvador, Finland, France, Germany, Greece, Honduras, Ireland, Israel, Italy, the Netherlands, Netherlands Antilles, Nigeria, Poland, Portugal, Thailand, Trinidad, West Indies, the United Arab Republic, the United Kingdom, and Viet Nam.

The ground-based program maximizes the scientific value of satellite programs by stimulating the gathering of important supplementary information and by greatly expanding the number of competent scientists attacking the sizable tasks of data analysis and correlation. Further, it engages foreign scientists in space-related activities, inspiring their continuing interest and imparting the knowledge necessary for further activity.

Such programs will be conducted whenever their potential scientific value warrants. The elements of these programs are sufficient early information to permit foreign participation, dispatch of current orbital information, and procedures for furnishing or exchanging data.

OPERATIONS OF OVERSEAS FACILITIES

NASA overseas tracking and communication stations, located in or near foreign communities, present a unique opportunity for contributions to the pattern of open co-operation in space research. Such contributions are assured by fostering greater operational participation by foreign nationals and by appropriate co-operation and service arrangements with independent foreign facilities in support of NASA programs.

Of more than two dozen overseas facilities, in eighteen different political areas, about two-thirds already operate wholly or in part with the assistance of foreign nationals. Tracking facilities under these various types of arrangements are to be found in Argentina, Australia, Bermuda, Canada, Canton Island, Chile, Curacao, Ecuador, India, Iran, Japan, Mexico, Nigeria, Peru, South Africa, Spain, the United Kingdom, and Zanzibar. Indeed, the cost of operating several of the stations is fully borne by the co-operating countries. Increased participation in the operation of the global network is encouraged, and a training program for this purpose is under way.

The specific contributions of the Jodrell Bank radio telescope and general interest in participating in United States satellite and space probe programs have created wide interest abroad in the establishment of new, advanced radio telescope programs, particularly in France, Italy, Japan, and Germany, while the existing advanced program in Britain appears likely to be extended.

To facilitate their future co-operation with NASA, agencies abroad are being provided by NASA with design considerations which may be taken into account in constructing their own facilities. These installations are certain to become important in future NASA programs.

TECHNICAL TRAINING

Scientific communities entering into the new technology of space research may find that their greatest need is for technical advice and experience. A postdoctoral program, with funds from NASA and administered by the National Academy of Sciences, makes it possible for foreign as well as domestic scientists to pursue space-connected projects in this country.

In a second and separate program, NASA offers laboratory support and training for extended periods to qualified scientists appropriately sponsored by their governments. The sponsoring government ordinarily meets travel and subsistence costs. Such laboratory support may be provided as part of a broader program in co-operation in space science. Eighteen nations are taking part in this program: Argentina, Australia, Canada, Chile, China, Denmark, Ecuador, West Germany, India, Israel, Italy, Japan, New Zealand, Norway, Pakistan, Sweden, Turkey, and the United Kingdom.

A multiplicity of training locations and fields allows the placement of a reasonable number of trainees without undue burden on any one office. The possible locations include Goddard Space Flight Center, Jet Propulsion Laboratory, Wallops Station, and Goldstone as well as other NASA field centers and university laboratories engaged in NASA programs. The possible fields include vehicle and launch operations; payload design, packaging and testing; space science programs and theory; tracking, telemetry, and communications; and data processing. Accordingly, assignment for training purposes is thus far proceeding on an individual basis and will continue to do so unless the number of trainees increases substantially.

More recently, NASA has established a third program—the NASA International University Program—to assist regional and national space research organizations abroad to develop scientists and technicians for their space research programs. Fellowships enable foreign graduate students to study and participate in research in the space sciences at leading United States universities for one to two years. University costs and necessary travel expenses within the United States are met by NASA. The sponsoring institutions abroad are expected to defray international travel costs and subsistence in the United States.

Supplementing the International University Program and in conjunction with co-operative program preparation, NASA is providing for scientists from United States universities and

NASA centers to go abroad for periods of one to twelve months to lecture to university groups on space research.

EXCHANGE OF INFORMATION

To ensure dissemination of scientific data resulting from space research, procedures are in force within NASA to provide for: dispatch of preliminary technical information to COSPAR upon the launching of rockets and satellites; regular transmittal of orbital elements and satellite observations through the international Spacewarn system designated for that purpose; NASA support of the United States component of Spacewarn; publication of preliminary scientific results and the deposit of results in the World Data Centers; agreements with experimenters to provide the results required; and publication, for world use, of telemetry calibrations where useful. These activities have as their background the exchange arrangements made during the IGY and continued since then. In space research, COSPAR has international responsibility for the World Data Centers while the National Academy of Sciences has domestic responsibility. NASA has been fully responsive to the recommendations of the scientific community on data exchange, as expressed by the Academy's Space Science Board.

This brief description of NASA's international co-operative activities should be indicative of a vigorous and productive program in support of national policy, statutory obligations, and operational requirements. Problems are, of course, encountered in all programs. The very long leadtime implicit in this most difficult of technologies means that co-operative projects can mature slowly at best. This fact is often compounded by the state of the art, by fund limitations, or by both. Time is required for a nation to decide to enter seriously into the new technology and to organize, finance, and plan for the conduct of space research.

These prerequisites to co-operation are being met by other nations. It is very probable that the NASA offer to COSPAR, made in March, 1959, through the National Academy of Sciences, was a significant factor encouraging the organization of national space committees in much of Europe and Japan, for it offered to foreign scientists an opportunity, not then existing or in prospect, to project their own experiments into the spatial environment.

Now other efforts to mount co-operative assaults upon the unknowns in space are materializing. Both in Europe and in Latin America, multinational regional organizations are perfecting their charters and considering appropriate programs. The United States has welcomed these developments and made clear its willingness to enter into joint projects with the new organizations on the same basis as with individual countries. Thus, the prospects for broad international co-operation in space research are multiplying.

It is the belief of NASA that international co-operative effort should extend to all nations, including the U.S.S.R. There is reason to believe that Soviet scientists desire to join their Western colleagues in useful and interesting enterprises. Overtures of co-operation, both general and specific, and provision of some co-operative service have thus far not developed into any tangible program, but it is hoped that a substantive start can soon be made. (Preliminary agreements on a bilateral program were reached in Geneva in June, 1962, in a discussion initiated by Mr. Kennedy.) Co-operative efforts by the rest of the world will in any case continue. NASA has found a warm and welcome response to its offers of co-operation.

THE UNITED NATIONS AND OUTER SPACE

CHRISTOPHER WRIGHT

The United Nations is our most comprehensive international political organization, and space science is one of the most complex sciences in terms of its social requirements. Therefore, it should not be surprising that each has taken notice of the other. Early in 1957 the United States proposed to the United Nations that disarmament arrangements governing research and development involving rockets and satellites should be worked out while the uses of outer space were still hypothetical or at most experimental. In November, 1957, six weeks after the Soviet launching of the first artificial earth satellite, the United Nations General Assembly passed, over Soviet objections, a resolution on reduction of armaments which provided, among other things, for "the joint study of an inspection system designed to ensure that the sending of objects through space shall be exclusively for peaceful and scientific purposes."

Until late 1958 the primary official interest in outer space was its connection with disarmament, with the Soviet Union linking the matter to the question of military bases on foreign territory. In the fall of 1958 the thirteenth General Assembly of the United Nations moved to consider various international threats and opportunities associated with space exploration. A resolution was passed to appoint an *ad hoc* committee of representatives of eighteen nations to review existing United Nations activities and resources relating to the peaceful uses of outer space, to explore areas of appropriate and possible international co-operation, to consider organizational arrangements for facilitating such activities within the framework of the United Nations, and to identify legal problems which might arise from activities associated with the exploration of outer space. The general form of this resolution agreed with a Soviet draft except that the composition of the committee differed substantially. As a result the Soviet Union together with the two other Soviet bloc members, Czechoslovakia and Poland, boycotted the committee, while two other members, India and the United Arab Republic, stayed away on the grounds that the committee could not function effectively without the participation of both space powers.

Despite the absence of five of its members, the committee prepared a report for the fourteenth General Assembly in the fall of 1959. One part of the report discusses legal problems which may arise in carrying out programs of space exploration, emphasizing the close relation between relevant law and space science and technology, and distinguishing between legal problems amenable to early treatment and those which are not yet ready for solution. Among the former are questions of overflight, liability for damage, radio frequency allocations, identification of space vehicles, and the re-entry and landing of space vehicles. Although no solutions are offered or mechanisms proposed for arriving at and enforcing solutions, this preliminary survey of problems, of the state of the law, and of priorities for official study is a prerequisite to a substantial contribution of the type which only the United Nations can make to the orderly exploration of space.

The *ad hoc* committee recommended that no new specialized agency or other autonomous intergovernmental organization be established for international co-operation in the field of outer space, and that no existing specialized agency assume over-all responsibility. It suggested that a unit of experts be established within the Secretariat of the United Nations to provide a center for facilitating international co-operation among the various governmental and non-governmental organizations undertaking space activities. It also suggested that the General Assembly establish its own committee to consider means for facilitating co-operation, and for studying and resolving legal problems.

This unit has not been established, but the General Assembly did create a committee on the peaceful uses of outer space, with twenty-four member nations for 1960 and 1961. This committee included the eighteen members of the *ad hoc* committee: Argentina, Australia, Belgium, Brazil, Canada, Czechoslavakia, France, India, Iran, Italy, Japan, Mexico, Poland, Sweden, the Soviet Union, the United Arab Republic, the United Kingdom, and the United States, together with Albania, Austria, Bulgaria, Hungary, Lebanon, and Rumania. The Soviet Union was in favor of this resolution. No other attempt was made to act upon the report of the *ad hoc* committee or to direct the new committee to do so.

The major differences between the directives for the two committees were that the new committee was not directed to consider the activities or resources of the United Nations or new organizational arrangements. The new committee was instructed to review again possible international co-operation, the legal problems arising from the exploration of outer space, and the means which could be taken under United Nations auspices for giving effect to national programs. In addition, the new committee was specifically requested to plan the international scientific conference on space exploration which the General Assembly decided to hold in 1960 or 1961. Despite this request there has been no planning, preparation, or scheduling of the conference. Indeed, the committee delayed meeting for the apparent reason that the Soviet Union and the United States disagreed on who should direct the work of the committee or the program of the international scientific conference, and even whether the committee should operate under normal voting procedures, as the United States insists, or should operate on the unprecedented basis of unanimity or a two-thirds vote as proposed by the Soviet Union.

The persistent deadlock between the two space powers promised the prospect of a permanently inactive and possibly even memberless committee. Thus, without referring to this committee the United States has indicated by various statements, including President Kennedy's address before the sixteenth General Assembly on September 25, 1961, that the United Nations should extend its charter to the limits of man's exploration in the universe; reserve outer space for peaceful purposes; prohibit weapons of mass destruction in space or on celestial bodies; and open the mysteries and benefits of space to every nation. President Kennedy promised United States proposals on these matters as well as on a global system of communications satellites and co-operative efforts between all nations in weather prediction and eventually in weather control. Subsequently the United States indicated that the General Assembly itself might implement these suggestions by explicitly confirming that the United Nations charter applies to the limits of space exploration, by declaring that space and heavenly bodies are not subject to claims of national sovereignty, by instituting an international system for registering all objects launched into space, and by creating a specialized space unit in the United Nations Secretariat.

INTERNATIONAL SCIENCE AND OUTER SPACE

Despite the barriers to action in the field of outer space within the United Nations itself, other international governmental organizations, and in particular three of the specialized agencies of the United Nations, are actively contributing to the exploration of outer space.

The United Nations Educational, Scientific, and Cultural Organization (UNESCO) has made indirect and primarily financial contributions to the planning of space explorations since before the launching of the first satellites. A 1951 agreement between UNESCO and the International Council of Scientific Unions (ICSU) provides for co-operation on scientific projects of mutual interest and for an annual subvention to facilitate co-operation and financing of international scientific projects conforming to the objectives of UNESCO. A special subvention assisted the planning of the International Geophysical Year (IGY). The annual subvention supports a major share of the ICSU budget and although the sums involved are modest compared to the costs of experiments in outer space, the importance of this kind of planning activity far outweighs its cost. The ICSU and its special committees act as clearinghouses by covering part of the costs of international meetings of scientists, by contributing to the planning of scientific activities, and by exchanging scientific information.

In recent years UNESCO has initiated efforts to strengthen mutual understanding and enhance the benefits which can result from this agreement. It proposed that there be closer relations between UNESCO and ICSU and resolved, by intergovernmental action, to stimulate research and promote appropriate international co-operation by such means as, for instance, facilitating the purchase of rockets. UNESCO now sends an observer to meetings of the ICSU Committee on Space Research (COSPAR) and plans to support specific projects of COSPAR and the International Astronautical Federation, as well as any projects connected with space exploration which the United Nations itself may undertake. It should be noted, however, that UNESCO selects specific areas of scientific research which it assists, on the assumption that the solution of these problems may help to improve the living conditions of mankind. For the period 1959–60, "exploration of extraterrestrial space" was last on a list of eight areas.

PROBLEMS IN COMMUNICATION

The International Telecommunications Union (ITU) is responsible for the international co-ordination and orderly use of radio and other forms of telecommunication. It maintains a registry of frequencies in use, and attempts to reconcile the assignments made by nations as well as to achieve equitable and effective reallocations or allocations of unused frequencies for specific purposes. The ITU has a scientific committee, the International Radio Consultative Committee (CCIR), concerned with the technical problem of making maximum use of the radio spectrum and avoiding interference. The committee seeks advice from the International Scientific Radio Union (URSI), a part of the ICSU.

Because of the capital investments involved, reassignments of radio frequencies are infrequently reviewed. Indeed, the most recent meeting of the administrative radio conference of the ITU prior to 1959 was in 1947, long before the exploration and exploitation of outer space were seriously contemplated by many people. In August, 1959, the Administrative Radio Conference of the ITU considered proposals by some of its members and the CCIR concerning frequency allocations needed for the exploration of outer space, including both space communication and radio astronomy. Apparently only as a result of last-minute efforts by interested American scientists did United States representatives to the ITU conference maintain a position compatible with the needs of the radio astronomers and space scientists. The agreements reached by the ITU in 1959 provisionally allocated one channel for radio astronomy (the hydrogen line at 1400 to 1427 megacycles per second) and thirteen channels for space communication other than between terrestrial stations via objects in space, a matter about which no decisions were reached. The thirteen bands cover 10, 20, 40, 136, 184, 400, 1428, 1700, 2300, 5250, and 8450 megacycles per second and 15.20 and 31.6 gigacycles per second (a gigacycle is 10^9 cycles). Already there are indications that these assignments will be inadequate as space exploration proceeds. In anticipation of expected rapid advances in space research, and despite the fact that only a very few of the 100 or more members of the ITU have direct interest in space exploration, an ITU Extraordinary Conference to consider frequency requirements associated with space activities is proposed for 1963. It is also expected that the next regular conference will be scheduled for 1965, thus reducing the period between such conferences.

The effectiveness of the ITU as a co-ordinator of communications systems depends upon its knowing the needs of member nations and allocating frequencies on a long-term basis with a minimum of inconvenience at any one time. When the allocations have been violated, which sometimes happens, there appears to be little that the ITU can now do to police its regulations. Space exploration promises to test the resilience of the ITU because of its need for specific frequencies possessing qualities favorable for research, because both the quantity and specific quality of its research and operational needs may change rapidly, and because reliable space communication may become so important as to require new means for protecting it from interference. In a few years there may be space vehicles traveling about the earth, moon, and solar system in such numbers that accidental or malicious interference with command signals and other telemetry will be a significant possibility, and a possible source of extreme national frustration and irritation.

To add to these problems, URSI includes among its leaders radio astronomers with their own professional interest in the outcome of certain ITU deliberations. H. C. van de Hulst notes (p. 291, below) that interested scientists have joined in an inter-union committee including URSI and are preparing a "scientific case" intended to assure that space science and radio astronomy will be protected and unhampered. It is hoped that it should still be possible for URSI to advise the ITU with the objectivity, if not the disinterest, of its "detached scientific approach to any radio problem including those that might in application have a political coloration," which was recognized by the U.N. *ad hoc* committee. Nevertheless, the many new considerations mentioned above must be expected to strain the co-ordinating procedures of the ITU and cause a re-examination of the type of international regulation which will be needed when space exploration stimulates increased amounts and varieties of communication through outer space.

METEOROLOGICAL SATELLITES

In contrast to the telecommunication requirements of space exploration and exploitation, the use of satellites for meteorological studies and routine weather reconnaissance on a global basis seems destined to present more new problems of international organization and operation than of regulation. The stated objectives of the World Meteorological Organization (WMO) are to facilitate world-wide co-operation in the establishment of networks of meteorological stations and to promote uniform standards and new uses for weather information. Ordinarily it acts as a clearinghouse for the exchange of information and the making of agreements among members, and not as an operating organization. It did, however, undertake responsibilities for some programs for the IGY. In 1959 the World Meteorological Congress of WMO resolved that WMO would encourage the development of satellites for meteorological purposes and would collaborate with other international governmental and non-governmental organizations to this end. It has an agreement with ICSU which facilitates close co-operation in several areas of inquiry stemming from the IGY, including the oceans, the Antarctic, and outer space.

It appears to be the view of WMO that the use of satellites for routine weather reconnaissance may have many desirable features but that it will create special operational problems which will at least require co-ordinated facilities for the reception, analysis, interpretation, and distribution of world-wide weather data. Thus R. C. Sutcliffe, writing in the WMO *Bulletin* (July, 1960) on co-operation between WMO and the International Association for Meteorology and Atmospheric Physics, referred to a greater intermingling of activity and less reliance on liaison. And in response to the 1961 report of its panel of experts on artificial satellites (which the panel's Russian member did not help prepare), the executive committee of WMO passed a resolution (Resolution 10CEC-XIII) requesting regional associations to study the possibility of establishing main or local read-out stations.

As H. Wexler and D. S. Johnson observe (see p. 7, above)—meteorological satellites will contribute strongly to the development of a single world meteorological system. Such a system and the prospect of its close ties to responsible international governmental agencies is, perhaps, foreshadowed by their observation that once meteorological satellite systems observe phenomena which may have serious consequences for the safety of people, it will be imperative that reliable means be developed for conveying such information speedily to all nations.

These brief descriptions of the role in outer space of the United Nations and its specialized agencies suggest that the prevalent view among governments is that the United Nations needs to take cognizance of the role and activities of non-governmental bodies organized by the international scientific community.

OUTLOOK FOR UNITED NATIONS CO-ORDINATION

Thus far the participation of the United Nations and other international groups in space exploration has been limited to solving problems of international co-operation and co-ordination, exchange of information and ideas, research design, and similar matters. In the background of present international governmental participation in space exploration are the events of the United Nations disarmament negotiations, the IGY, and the Geneva atoms-for-peace conferences. Seen in retrospect, these events dramatize the existence of major policy issues having to do with the relations between arms control and scientific progress, between national and international scientific activities, and between diplomatic and scientific initiatives on the international level.

The international disarmament and arms control negotiations carried on over the past fifteen years may have served many useful purposes but thus far they have not achieved their stated objectives. International discussions and plans concerning the uses of outer space might have remained in this context of concern about arms. The fact that they appear not to have done so, as demonstrated by the terms of reference of the two committees on the peaceful uses of outer space and the report of the first, as well as by the proposals for United Nations action put forth by the United States, appears to be due to a realization that technical military applications built upon new sciences are not easily arrested and that an effective way to control the arms race may be to approve or even encourage open space explorations when it appears at least as likely to have peaceful as nonpeaceful implications. Plans for international explorations concerned with nuclear physics and atomic energy did not receive such attention at a comparable period in history. The emergence in the United Nations and its specialized agencies of this more flexible approach to major scientific and technical advances may well have been due in part to the positive successes of the IGY and the Geneva conferences on the peaceful uses of atomic energy as well as to the inconclusiveness of negotiations on general or specific arms limitations.

The success of the IGY has already been established both as a scientific venture and as a demonstration that international co-operation on scientific matters is possible if, and perhaps only if, scientists are permitted to co-ordinate plans and operate on a world-wide but non-governmental basis as they did in this case through a special committee, The Comité Spécial de L'Année Géophysique Internationale (CSAGI), of ICSU. The organization of the IGY placed emphasis upon international co-ordination and co-operation by scientific groups with virtually all financing and operations conducted either privately or officially by participating nations. Regardless of how closely these groups are affiliated with their governments they tend to favor organizations with proven ability to facilitate specific technical tasks and minimum bureaucracy. With this precedent of the IGY in mind the United States and other nations took the position that the United Nations should not itself undertake to plan and operate a program of space exploration, despite agreement within the United Nations that space exploration should be furthered for the benefit of mankind and of all nations and not be the means for national advantage or rivalries.

A resolution of the fourteenth General Assembly of the United Nations pointed toward a conference on space science and technology similar to the international conferences on the peaceful uses of atomic energy, except that it made no specific mention of participation by individual experts in peaceful uses of outer space, excluded participation by non-members of the United Nations, and gave planning responsibility to the U.N. committee on the peaceful uses of outer space. Thus the Secretary General and his advisory group are not in a position to implement the decision to have an international scientific conference on outer space as they were in the case of atomic energy conferences.

The findings of the *ad hoc* committee on the peaceful uses of outer space indicate what appears to be the general belief that the structure and actual history of the International Atomic Energy Agency (IAEA) do not provide satisfactory guidelines for promoting the peaceful uses of outer space. The IAEA was established in 1956 for the dual purpose of realizing the aspirations of nations wishing to harness the atom for peaceful purposes and of maintaining or even extending international security against the development of atomic weapons. Although it has powers of inspection and sanction which can be used in connection with its activities, it has suffered from uncertainty about its proper role and from the disproportion between the demands of small-nation members and the support, or lack of it, given by the major nuclear powers. With space exploitation as with atomic energy, the difficulty of separating features affecting national security and bilateral relations from those which are solely scientific or peaceful, together with the cost of the resources involved, makes general international operations in these fields rather unsatisfactory.

As an alternative, United Nations resolutions concerning outer space call for continuation of space research within the framework of the IGY. Either directly or indirectly, the *ad hoc* committee, UNESCO, WMO, and the ITU have indicated the importance of COSPAR as an international non-governmental scientific organization in the area of space science. It remains a question, however, whether the vocal support of COSPAR represents a conviction that the IGY mode of operation can continue, or a reluctance to develop any other frameworks in case this mode becomes ineffective.

It would appear that the events described here have influenced the character of present interest in space exploration within the United Nations to the extent that the idea of space exploration has not been as closely tied to the issue of arms control and national security as it might have been. These events have also demonstrated that international organizations can contribute to space exploration and that initiative by scientists operating in an international context can serve a vital purpose. For reasons indicated below, these issues are by no means settled. Arms control debates in connection with space activities may take on new force; international initiative in space matters may be further replaced by national or bilateral arrangements; and the increased acquiescence to leadership by the scientific community may be checked by the presence in scientific conferences and organizations of governmental representatives operating under political direction.

POLITICS AND PROBLEMS

There are indications that the interaction of international politics and the exploration of outer space will produce an increasing number of problems related to political parity, the identification of peaceful uses of outer space, and the location of space exploration among scientific and other human enterprises.

Political parity between the Soviet Union and the United States has been a long-standing objective of Soviet policy. The United States program for launching a manned lunar expedition is in large part a recent parallel response to this objective. The intrusion of political parity into the affairs of science is evidenced in the United Nations by the disputes over the composition of the committees on the peaceful uses of outer space and the direction of the proposed scientific conference. The new committee gives the Soviet bloc proportionately greater representation than it had on the *ad hoc* committee and, through changes in the decision-making procedure, the Soviets expect to have complete parity. The Soviet view is that the conference must reflect parity on space matters and also serve as a counterbalance to United States initiatives in the earlier conferences on the peaceful uses of atomic energy. Special veto powers for the Soviet Union and the United States are now built into the COSPAR charter even though the scientific effectiveness of such committees rests essentially on a spirit of voluntary co-operation. The management of other international non-governmental technical groups is also reflecting the idea of political parity. One cannot escape the fact that the popular attention and practical achievement of the atoms-for-peace conferences and the IGY proved them to be effective platforms for national comparisons, thus making the planning of further international science activities a matter of increasing political, which also means bureaucratic, interest even though these initial successes were in part due to this very lack of major political interest.

Parity in this context points to a "space race," if not a "science race," and implicity assumes that the relative standing of two nations in scientific and technical matters can be formalized. Although the present incursion of international politics into science singles out space explorations for political reasons and assigns standards of success which may not, themselves, promote co-operation, a space race may, for instance, offer innocuous means for demonstrating changes in potential military strength and could facilitate non-violent adjustments. Support of space projects or of any other scientific efforts is quite likely to generate unexpected discoveries which will frustrate efforts to establish influential international political bodies to deal with science and technology.

Peace is a political concept and the pursuit of peaceful activities a desired human goal. Statesmen have a responsibility to create overt or tacit standards, controls, and regulations where these seem necessary to minimize human disputes and to settle those that do arise. Whether one takes a broad or narrow interpretation of the meaning of "non-peaceful uses," present conditions of military technology and strategy indicate that not all basic research in space science or related subjects is likely to be compatible with practical arrangements ensuring that outer space is used only for peaceful purposes.

Complementary to the international political interest in keeping the peace, even at the expense of space science, is the widespread human aspiration to increase the sum of scientific knowledge and enhance respect for the objective universality of scientific inquiry. These interests intrude upon international politics and point to the need for political decisions within science. Space exploration is only one of many opportunities for legitimate scientific inquiry—opportunities which together tend to saturate the resources and attention capable of being given to science. Questions of competition and co-operation among sciences and their practitioners can no longer be ignored or allowed to solve themselves in the passage of time. Efforts will have to be made individually, nationally, and internationally to estimate objectively the likely contributions of various scientific enterprises to the advancement of mankind and to act upon these estimates. The continuing critical attention given to space exploration and other topics in the United Nations and its specialized agencies may help scientists and non-scientists alike to decide for themselves the priority which space exploration deserves.

COSPAR AND SPACE CO-OPERATION

H. C. VAN DE HULST

Scientific research is difficult. It is hard to do it alone; it is even harder to do it together. Any readers may know by their own experience that science is akin to art, that it requires a concentration which at times excludes consultation, that its emotional drive may be lost when channelized too strongly in projects or agreed divisions of tasks, and that a search may sometimes end in a surprising discovery which makes the method followed seem ridiculous, whereas another equally promising search may end in mere disappointment. They also know how much caution is required, just because of these circumstances, in discussing unfinished investigations, how many checks and double-checks are needed before a result is "established." They know how this attitude is stronger in one individual than in another, and in one group of research workers than another—and how difficult it may be to distinguish this cautious silence from secrecy imposed for security reasons.

All of this sounds rather old-fashioned and ivory-tower-like in the century of efficiency, teamwork, and projects, and in a time when even awarding Nobel Prizes becomes more and more difficult because it is hard to tell "who did the work." Most scientists will agree that co-operation in scientific research between individuals or groups requires *mutual confidence* first, *mutual understanding* in the second place, and *efficient organization* only in the third place.

The present international scientific unions grew from the necessity for exchange of information by means of personal contacts. They started, as early as 1919, as groups of colleagues who knew each other by visits, correspondence, and by studying each other's scientific publications. Their gatherings intensified such contacts. A certain standardization in terminology and methods was achieved, and moral and financial support was often given to the continuation or publication of important work. Prospects and arrangements for future research were intensively discussed. This was and indeed still is, a quite efficient form of international co-operation in science. Many international co-operative plans have been conceived in this manner, and perhaps equally important, much waste of effort avoided, when immature plans were canceled or modified by the advice of wiser colleagues.

It is necessary to spell this out because there is a world-wide tendency for progressively increasing government initiative and control in many ways of life, including science. In some countries this is a professed policy, in others, an unavoidable development. Many people, in many countries, believe that truly international organization in science should be achieved at the level of contacts between governments or government agencies. They assume that international co-operation in science always starts with a framework agreed between governments; the scientific details may then be left to be filled in by the professional scientists.

The weakness of either standpoint is evident. The traditional approach by means of friendly contacts between individual scientists contains a strong guarantee against schemes which are scientifically empty or worthless. But the scientists can seldom commit their nations' institutions to carry out part of a proposed plan. Usually a period of persuasion has to follow, in which resolutions passed by a (non-governmental) international scientific union may be handled as important but by no means conclusive arguments.

Direct agreements between governments, on the other hand, provide a sufficient guard against plans without commitments but are in grave danger of leading to schemes which the scientists will find useless or impossible.

HOW COSPAR IS ORGANIZED

The Committee on Space Research, (COSPAR), was confronted from its start with these dual attitudes and the solutions found are interesting. Established as a special committee of the

COSPAR and Space Co-operation

International Council of Scientific Unions (ICSU) in November, 1958, it has functioned under its present charter since December, 1959. The full committee is formed by the representatives of ten (out of a total of thirteen) international scientific unions—the international unions of astronomy, of biochemistry, of biological sciences, of mathematics, of theoretical and applied mechanics, of pure and applied physics, of pure and applied chemistry, of physiological sciences, of scientific radio, and of geodesy and geophysics; and eighteen of the fifty national academies or similar institutions which form the "national" membership of ICSU—Argentina, Australia, Belgium, Canada, Czechoslovakia, France, the German Federal Republic, India, Italy, Japan, the Netherlands, Norway, Poland, South Africa, Switzerland, the United Kingdom, the United States, and the Soviet Union.

ICSU had also sponsored the International Geophysical Year, and COSPAR, established shortly after this year ended, may be regarded as a direct successor to one of the fields of activity of the International Geophysical Year. The fact that COSPAR is a daughter organization of ICSU has lent it a certain stability, both in political and in administrative matters, from its difficult start.

The two types of representation, copied directly from ICSU, form in my view a good solution to the problems mentioned above: The international scientific unions may judge world science in their field, ideally with disregard of any politics. The national academies have sufficient standing in their own country to make commitments about joint plans and yet they stand somewhat (or entirely) apart from government politics and can select their representatives on the basis of scientific competence only.

COSPAR has made its executive council, which consists of the representatives of the international scientific unions, responsible for the formulation of the programs and policies, while these same representatives are excluded from the vote on any practical plans involving major financial commitments of the countries. The parity question, which frustrated the work of the United Nations committee and which delayed the work of COSPAR during 1959, was solved by a bureau consisting of a president (H. C. van de Hulst); a vice-president from the United States (Richard Porter); and a vice-president from the Soviet Union

(A. A. Blagonravov); and four members, two elected from a slate submitted by one vice-president and two from a slate submitted by the other vice-president—E. Buchar (Czechoslovakia); H. S. W. Massey (United Kingdom); Maurice Roy (International Union of Theoretical and Applied Mechanics); and W. Zonn (Poland). This gives due recognition to the two countries far advanced in space research, namely, the United States and the Soviet Union, but wisely it does not recognize any blocs of countries such as exist in the political field. The composition of the bureau shows indeed that the president and one further member are representatives of international unions and not national representatives. Incidentally, the customary officer called "general secretary" did not fit into this solution. An executive secretary, reporting directly to the president, was appointed to direct the secretariat.

HOW COSPAR WORKS

COSPAR has followed very much the pattern of work established by the international scientific unions—occasional large meetings are devoted both to scientific papers and to business sessions. One such meeting was held in Nice in January, 1960; another in Florence in April, 1961, the third in Washington, D.C. in May, 1962. Between these meetings there may be a few committee or working group meetings and exchange of information by correspondence.

The international scientific unions have fields with traditional borderlines, not always logical but fairly well established. COSPAR cuts across these boundaries to deal with "scientific research in the widest sense, carried out by means of rockets or rocket propelled vehicles." While it might seem that COSPAR must always infringe on the fields belonging to the unions, in practice there are so many new techniques, new branches and interfields, that there is plenty of choice. Since most of these unions have had an enormous increase in the number of persons —and number of countries—doing active research in their fields and wishing to attend the meetings, the problem is how to prevent a meeting of a thousand or more scientists from becoming a succession of advanced popular lectures, finished by a plenary session applauding resolutions dictated by a small governing board. Or, more positively: How is it possible to recapture the spirit of thorough personal discussion of a field of common interest which characterized the earlier small meetings? The answer is generally sought—and with fair success—in the direction of small, highly specialized meetings, sometimes even without the presentation of formal papers.

COSPAR tried to follow this. The Nice symposium in 1960 was a symposium on space science in general. This meant already a considerable restriction compared with the topics which are usually gathered under the term "astronautics," for it excludes everything about the technique of the rocket vehicle and about manned space flight. Nevertheless it was judged useful to restrict the Florence symposium even further—the Washington symposium was almost unavoidably a large one but plans are to make the Warsaw meeting more specialized.

ACHIEVEMENTS OF THE COMMITTEE

What has COSPAR achieved? It is difficult to answer this question objectively, for there is no standard measure to assess the results. One might mention the precise numbers of circulars, letters, and telegrams mailed by the secretariat. Among the circulars have been a number of advance notices of such technical details of planned launchings—as yet only from the United States—which may enable scientists in other countries to prepare the relevant observations. Other information, also about space research meetings which were not sponsored by COSPAR, such as the regional conference in Buenos Aires, October, 1960, is contained in the COSPAR *Bulletin,* which has appeared quarterly. One might also point to the impressive symposium volumes —titled *Space Research,* published in Amsterdam in 1960—each containing over 100 papers, all with abstracts in Russian and in one of the Western languages. (See references, p. 298.) This is indeed an achievement; nevertheless, a skeptic would do well in describing such a volume as mainly a tangible proof of a certain willingness to co-operate, rather than new results made available to the scientific world by COSPAR's efforts. For it is clear that the majority of these results would have been published eventually in the scientific literature.

Yet the real achievements are imponderable. On the eve of the Nice symposium, an American and a Soviet scientist, who had not met before, spent three hours editing the English abstract of the latter's paper. Evidently, this was not only a linguistic exercise but also a comparison of terminology, unstated assumptions, and scales of importance. I tend to regard these hours as far more efficiently spent for the cause of international co-operation in science than many hours of formal lecturing, although these are also necessary.

Active co-operation in the form of joint experiments, initiated or sponsored by COSPAR, has not gone very far yet. The Rocket Weeks and Rocket Intervals agreed by COSPAR have been put on the International Geophysical calendar and have indeed been used by several countries for almost simultaneous experiments. Telemetry codes of some space experiments have been released by the American sponsors and distributed by COSPAR. Some other plans are not beyond the stage of recommendations. But again, the actual credit may be higher, for many co-operative plans will not finally take shape within the COSPAR organization. It is the explicit policy of ICSU's special committees to encourage informal discussions about bilateral arrangements but to refrain from recommendations, in order to leave the decision fully to the countries or institutions involved.

COSPAR AND THE U.N.

Finally, the relation between COSPAR and the United Nations organization and its specialized agencies is worth mentioning. Since these bodies are governmental and COSPAR is non-governmental, there cannot be an organizational relation. However, there are plenty of contacts and, in general, a good understanding of COSPAR's task by the bodies concerned. COSPAR has offered services to the U.N. in case the committee on the peaceful uses of outer space might wish to use such services. Relations with UNESCO and with the World Meteorological Organi-

COSPAR and Space Co-operation

zation are in a sense prescribed by the agreements which ICSU has with these bodies. Observers of these and of the International Telecommunications Union (ITU) come to the COSPAR meetings and there is mutual interest in future plans. UNESCO gave financial assistance to the last COSPAR meeting.*

Great interest must be attached to matters of regulation or, possibly, legislation. COSPAR's task in this matter is described in its charter as follows:

Recognizing the need for international regulation and discussion of certain aspects of satellite and space probe programs, COSPAR shall keep itself informed of United Nations or other international activities in this field, to assure that maximum advantage is accorded international space science research through such regulations and to make recommendations relative to matters of planning and regulation that may effect the optimum program of scientific research.

Action under this clause was required when in August–December, 1959, the ITU held its ordinary administrative conference to review, among other points, the allocations of bands in the frequency spectrum to many classes of users. COSPAR was admitted in the status of observer and spent considerable effort in trying to emphasize in committees and in plenary sessions the need for open frequency bands for space science. At the same time, the International Astronomical Union and the International Scientific Radio Union (URSI) fought for free bands for a related (but not identical!) branch of science, radio astronomy. In either case it is a matter of life or death for these sciences, for a frequency spectrum entirely taken by other users, however useful, would mean that there would be no detectable signals from outer space. For science this situation was without precedent. Usually it has been possible to find "quiet places" for sensitive experiments by going far away from the cities; but this is only of very limited help in the present situation. Also for the ITU, the situation was without precedent: the observers on behalf of radio astronomy and space science were present during most of the three-month period and were often asked to speak and participate in committees and subcommittees.

---

*June, 1962—The U.N. committee on the peaceful uses of outer space has been active and has recognized COSPAR's role more fully than I dared hope.

A skeptic may again ask: what good are the results, a list of bands reserved with plenty of restrictions, for space science and radio astronomy? Has the voice of science indeed affected the heavy administrative machinery, in which strong political and commercial tensions determine the equilibrium? I think the answer is yes. Many national administrations appear to have an eye open both for the importance of science in its own right and for science as a necessary step toward later applications. Hence also this aspect of COSPAR's work must continue. The International Scientific Radio Union, the International Astronomical Union, and COSPAR have now joined forces in an interunion committee on the allocation of frequencies under ICSU. One of the first tasks of this committee will be to prepare the scientific case for the next ITU conference in 1963.

SUGGESTED READING

CIRA 1961: COSPAR International Reference Atmosphere. H. K. Kallmann, editor. Amsterdam: North-Holland Publishing Co., 1961.

Space Research. H. K. Kallmann, editor. Amsterdam: North-Holland Publishing Co., 1960.

Space Research II. H. C. van de Hulst, A. F. Moore, and C. de Jager, editors. Amsterdam: North-Holland Publishing Co., 1961.

PART FIVE

Space Technology

Introduction
Space Vehicles
Chemical and Nuclear Rocket Propulsion
Deep Space Propulsion Systems

INTRODUCTION

HUGH ODISHAW

Capability in outer space depends upon propulsion systems. This dependence is simple and clear, and might be compared to the dependence of oceanic commerce and research upon sea-going vessels. Accordingly, this section deals with propulsion systems: the chemical (liquid and solid) propellant engines, the development of nuclear engines, and the development of special, low-thrust devices that can play an important role in space once large engines have fulfilled their launching function. Some attention is also given to the auxiliary technology associated with space systems, such as guidance and control, and the implications of technological advances on space systems to other areas of technology are touched upon.

The emphasis in the following chapters is on propulsion because high-thrust engines determine the extent of space payloads and missions. But just as the sail or a marine diesel engine is only part of a ship, rocket engines are only part of the technological complex of a given space system. The engines themselves link together such components as the propellants and their containers, devices for channeling and controlling the flow of fuel and oxidizer (in liquid rockets), the combustion chamber, and the exhaust nozzle. Efficiency dictates two or more stages for such vehicles, but this requires discarding units that have performed their task, and controlled separation devices are needed. All the components, including payload, must be contained in a suitable housing, which must be light in weight yet rigid and strong as well as aerodynamically satisfactory. But more is required because space systems must carry devices suitable for guidance, control, stabilization and, on occasion, orientation.

The system, whether an artificial satellite or a deep-space probe, must be tracked. Spacecraft carry radio transmitters whose signals are intercepted by ground stations; computations on high-speed computers provide trajectory and orbital data. The electronic tracking techniques for earth satellites are based, in effect, upon a radio triangulation method, but optical photographic methods are also used and provide the best orbital data. For craft reaching far out into space the tools and techniques of radio astronomy are necessary.

Devices and techniques for recovery of payloads represent another large area of technical activity. Crucial to manned ventures is the safe return of the capsule. Instrumented capsules have also been recovered, in water and in air, by aircraft interception during the decline of a satellite, while simple balloon descent of sounding-rocket payloads has been used. Scientific data generally are acquired by telemetry: the observations of instruments are converted into coded electrical signals; these are transmitted to earth by radio, where they are decoded and analyzed with the help of computing machines.

Launching, too, entails a host of technical facilities and skills, as the complex of facilities at Cape Canaveral suggests—acres of launch pads and towers, networks of underground storage tanks and pipe lines, and control centers characterized by vast arrays of electrical and electronic instruments.

The very conduct of scientific experiments is intricately dependent on technology. No longer is the design of an experiment a laboratory affair, for the equipment must withstand the space environment as well as the acceleration and vibration during the launching phase, and it must be fitted into the total payload, linked to power supplies and to transmitters.

In short, the totality of all those aspects that are suggested by the word "technology" ranges far beyond the rocket engines, and even beyond such applications as communications and weather satellites. Moreover, the engineering and applied research devoted to fuels, to metals and ceramics, to structures, to engines, to electronic devices for guidance, stabilization, control and telemetry, to tracking and position computation, to data processing and analysis—these technological developments provide new materials, processes, devices and techniques that have applicability, beyond space technology, in other areas of technology, industry, and science itself. The sum of these is virtually a new

Introduction

technology and industry, rivaling in quantity the total economies of sizable nations. This complex has a meaning not confined to space achievements alone, but has an impact on society as a whole, if for no other reason than that the amount of effort, realized and potential, is vast.

SPACE VEHICLES

GEORGE P. SUTTON

Space system components, vehicles, launching and service systems, and techniques for integrating all systems, make up what might be called a space systems technology. This technology is large and complex, ranging from structural engineering to sensitive guidance and control devices, from component reliability and statistical quality control problems of types that are essentially understood to fundamental research on combustion, materials, or fuels.

The current state and the future development of this technology are significant in two ways: First, this technology is the means by which space achievements—research or applications—are realized, for it provides the vehicles, controls, and data links for space flight. Second, it represents a subject which has engineering, technological, and industrial significance, because its developments are applicable to areas other than space systems themselves. I shall outline some of the applications in other fields, while attempting to suggest the nature of the technology, some of the trends in the field, and some indication of its general impact.

VEHICLES

In order to attain the high velocities required to put payloads into space orbits (above 25,000 feet per second) or into our earth escape trajectory (above 37,000 feet per second) it is necessary to outfit the vehicles with high performance rocket engines (see Table 2, below) to minimize the weight of the empty vehicle (without rocket fuel) and thus to minimize structural hardware and to provide for an orderly control of the flight so that it will proceed along the intended optimum trajectory. Thus the story of space vehicles must be concerned with the systems that propel them, with optimum multistage lightweight structures (which weigh only about 10 per cent of the total takeoff weight of any given stage—the remaining 90 per cent being rocket fuel) and with reliable but very lightweight equipment for guidance, communications, instrumentation, and other functions.

Space launching vehicles used today draw largely on engineering experience obtained in military ballistic missile programs; the first space launchings were started as simple extensions of this technology. Some of the principal types and some of their characteristics are shown in Table 1. Only recently have we started to develop vehicles specifically designed for space flight which are different from the basic rockets and launching devices used for ballistic missiles. Centaur, Saturn, and future larger vehicles fall into this class.

The performance of these launching vehicles determines the capability and payload capacity of various space flight missions. One trend has been obvious: the vehicle sizes and payloads have steadily increased and are still increasing. Table 1 also indicates typical payloads, first flight schedule, and typical applications.

If we assume that some of the uncertain design problems of space flight will soon be solved (such as the radiation hazard and the effect of meteorite impacts), it is safe to predict that future larger vehicles will be developed for manned missions to the moon and to the planets beyond. Such larger vehicles are an absolute necessity for manned planetary space flight, and the development of a launching vehicle of a gross weight of more than 10 million pounds at takeoff is technically feasible today. They would be capable of carrying payloads in excess of 100,000 pounds to the moon, which is necessary for long missions with several human occupants.

Because we do not expect to have a very efficient propulsion system, we are forced to continue using multistage vehicles which have boosters, one or more sustainers, and a terminal stage. Each propulsion stage is designed and constructed differently from the others, and is optimized for its specific role in the mission trajectory. Thus it is possible to drop off the dead or useless weight of a stage whose propellant has been expended; no further energy is spent on accelerating this useless weight, and a very high mass ratio (weight of final stage after burnout divided by initial takeoff weight) is obtained by using several stages. Constructing space vehicles by assembling independent modules will find application in future versions; modular capsules for example, will be attached for specific purposes, augmented propulsion capability, additional communications capability, etc.) and will permit the development of integrated, complex space vehicle systems such as space stations.

The maximum payload that can be carried by any one vehicle is dependent on the difficulty of the mission, the staging, the exact mission trajectory, and the propulsion application-time

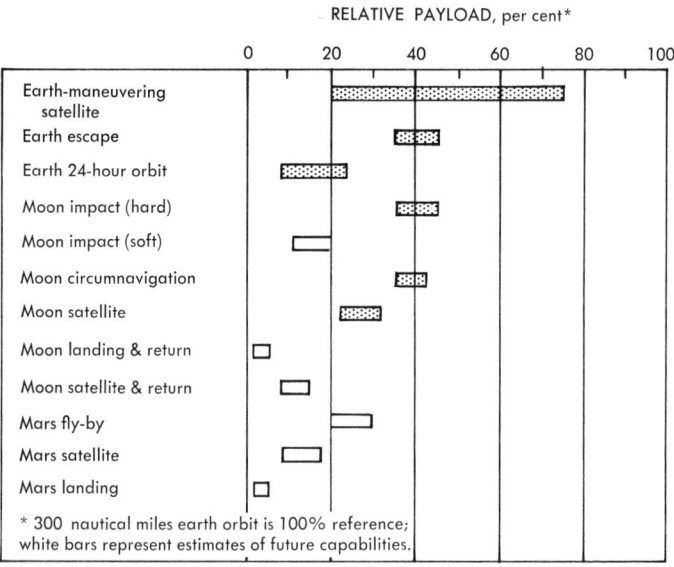

FIG. 1—Payload-mission conversion chart for high-energy chemical multi-stage rocket vehicles.

profile. A rough estimate of the tradeoffs between mission capability and payload is given in Figure 1. For example, the Atlas Centaur, which has a 300-mile orbital payload of about 8,500 pounds would, according to Figure 1, put about 35–40 per cent (or 3,000 pounds) of this into an escape orbit, or 12 per cent (or about 1,000 pounds) into a stationary 24-hour earth orbit.

STRUCTURES

Several novel structural approaches are being pursued actively to improve space vehicles and further reduce the weight of the empty vehicles. Sandwich and honeycomb construction of structures is being further refined to improve bending and thermal characteristics. One new idea is a parachute-glider combination, which is really a foldable and controllable flexible wing which provides very high-lift booster recovery. Another is an inflatable temperature-resistant structure, which might permit wings on re-entry vehicles to be inflated when needed, or be used for possible construction of future toroid-shaped space stations.

The pressure-stabilized shell of the Atlas-type vehicle has made it possible to develop a lightweight vehicle whose structure accounts for a low percentage of its weight. The use of high-pressure vessels for solid propellant rockets and for liquid propellant rockets with pressurized feed systems has made it necessary to develop to a high degree the technique of making low weight, high strength pressure vessels. Heat-treatable, high strength titanium sheeting for this purpose has been developed recently. In addition, filament-wound cases using glass, beryllium, steel, or titanium wires with a ductile matrix have been developed for high strength pressure vessels. This filament-wound tank technology is already finding its way into the chemical and petroleum industry.

Usually the most expensive parts of vehicle technology are the ground-support equipment and support systems necessary to make the vehicle work. These include assembly and checkout devices for every major subsystem in the vehicle and payload—launching facilities, range instrumentation, fixed and mobile (truck-mounted or shipboard) telemetering receivers, booster or vehicle recovery devices (for example, the Discoverer has a fleet of special aircraft to catch a re-entry capsule in the atmosphere while it is parachuting down), vehicle transporters and erectors, propellant fueling, handling and storage equipment, and tracking stations. Because some space vehicles are large (see Table 1) the problems of shipping, assembling, checkout, and

TABLE 1

TYPICAL SPACE VEHICLES

VEHICLE	FIRST FLIGHT	DIAMETER	TAKEOFF THRUST lbs.	NUMBER OF STAGES	PAYLOAD (lbs., 300-mile orbit)
Sputniks II, III, IV	Nov. 3, 1957	Unknown	600,000*	2 to 3	2,925
Thor Agena	Feb. 28, 1959	8 ft.	150,000	2	1,600
Thor-Able	April, 1960	8.5 ft.	150,000	3	200
Atlas-Agena	1960	120 in.	360,000	2	4,000
Atlas Centaur	1962	120 in.	360,000	2	8,500
Sputnik V	Aug. 19, 1960	Unknown	600,000*	2 to 3	10,000
Saturn C-1	1961–63	22 ft.	1,500,000†	2	20,000

* Estimated value.   † Later versions using 188,000-lb. thrust engines.

launch assume new dimensions requiring shipment of vehicles by water in special barges, and the construction of special launch towers. Improvements in diagnostic testing and checkout procedures for vehicle and payload subsystems have reduced the complexity of ground-based checkout equipment at the launch stand. The possibility of having to handle some nuclear subassembly (for example, for auxiliary power) by remote means (for example, by quickly disconnectable fittings), may in the future complicate ground assembly and checkout.

PROPULSION

To date, almost all space vehicles have used liquid propellant rockets for their primary motive energy source. Different propellants (liquid oxygen and jet fuel, nitrogen oxides and hydrazine-type compounds) are in use or are being developed, with higher energy propellants (such as liquid hydrogen with liquid oxygen) favored for the upper stages. The largest booster vehicle currently under development, the Saturn, uses a cluster of eight individual large rocket engines to give a takeoff thrust of 1.5 million pounds. Also under development is a single rocket engine (the F-1) which will have 1.5 million pounds thrust at sea level, and, when clustered, should provide the propulsive mechanism for future large interplanetary space vehicles. Solid propellant chemical rockets are used in several upper stages and in the Scout vehicle; new, improved, and larger versions are currently under study or development.

Although liquid and solid propellant rockets are now the only feasible and practical methods of propulsion for space vehicles, there are several other methods in the research stage, which may provide improved performance in the near future. They use principles other than chemical reaction for energy release, and offer higher performances as seen in Table 2. The successes of

TABLE 2

CHARACTERISTICS OF PROPULSION SYSTEMS

PROPULSION SYSTEM	SPECIFIC IMPULSE*	THRUST TO WEIGHT RATIO	TYPICAL DURATION
Chemical rockets	200 to 430	0.01 to 100	minutes
Nuclear rockets	500 to 1,100	0.01 to 10	minutes to hours
Electric systems	1,000 to 30,000	10^{-2} to 10^{-6}	weeks

*Pounds of thrust per pound of propellant flow per second, at 1000 psi chamber pressure and sea level.

the Kiwi series of tests on the operating characteristics of very high temperature, open cycle, gas-cooled reactors has encouraged the future development of nuclear rocket engines. These experiments have shown that materials, controls, feed systems, and reactors necessary for nuclear rocket propulsion can be developed and the launching of a nuclear rocket could take place in the latter part of the 1960's.

Various concepts for electrostatically accelerating charged particles, or for electromagnetically or thermodynamically accelerating plasmas, have favored the development of several types of electrical propulsion systems. All of these, of course, require a heavy, high-output, space-borne energy source, which will probably be a nuclear power reactor. These electrical propulsion systems provide low thrust, but they have the advantage of high performance and thus a relatively small rate of fuel consumption. Research has progressed sufficiently with these electrical propulsion devices to attempt several experimental space flights in the near future. These flights will test such phenomena as the neutralization of the ion beam in free space.

For optimum performance, future space vehicles will consist of multistage devices with different types of rocket propulsion systems in the several stages. It is very likely that for takeoff and landing, chemical rockets will be preferred for a long time to come. For interplanetary or space maneuvers, chemicals will compete in the future with both nuclear and electrical propulsion types.

POWER SOURCES AND CONVERSION

Every space vehicle needs an internal power supply for operating its instruments and tools, for transmitting its information, for the operation of its guidance and navigation system, for the comfort and survival of a crew, and for many other purposes. The power source and its conversion equipment will vary depending on the size of the power demand (actually the size of the payload package and its design) and the duration of the flight.

All power supplies for space vehicles seem to have to meet a power demand that is occasionally high, but has a low level, long duration drain; thus all these power systems must be flexible enough to provide for rapidly varying power outputs during their lifetime. A standby or spare power system is often used in space vehicles.

The technology and science of these power sources and conversion devices encompasses many fields, including thermodynamics, solid-state physics and petroleum chemistry. There are basically only three energy sources: chemical (such as a fuel cell or a positive displacement engine), nuclear (fission reactor or radioactive isotope decay), and solar radiation (solar cell or solar-thermal engine); however, there are many conversion mechanisms for transforming this energy into the most useful form (usually into electricity). The construction of each of these sources and their conversion devices has progressed to a different stage of development. Most of these power conversion devices are relatively inefficient, and much work needs to be done to improve their available energy output. Unfortunately many of the conversion devices are heat machines, and their efficiency is limited by the Carnot cycle efficiency.

Lightweight batteries which are used extensively for the current space vehicle effort will continue to be used for short duration, low-power applications. Other means of generating power chemically with fuel cells will continue to be refined and very likely will be applied to space payloads. Solar energy sources are generally effective for longer durations, although they are still relatively high in weight per unit of power output; because of their simplicity, they tend to be reliable. Current vehicles make extensive use of solar cells which use a photovoltaic principle for converting the energy of the Sun's photons into electrical energy.

Solar cells are semiconductors; when subjected to bombardment by high speed electrons and protons, they suffer and their performance is degraded. A protecting glass cover reduces this degradation, but causes an undesirably large increase in weight. Photovoltaic devices have no moving parts, are simple, and quite reliable. Because solar cells are not exposed to the sun when an earth satellite is in the shadow of our planet, it is necessary to provide an auxiliary energy storage system (usually batteries) together with an appropriate switching system. The thermal energy contained in solar radiation can also be used in a thermal cycle engine by concentrating the sun's radiation through an optical system on a receiver; this will be possible at higher power levels.

A small (4.4 pound) nuclear power source of 3.0 watts, which uses the decay energy of plutonium 238 and a thermocouple type heat-to-electricity direct power conversion device, was successfully flown for the first time in a Transit satellite in late 1961. For long duration, high-power levels, a nuclear reactor seems essential, and either a turbo-electric or a thermal-electric conversion method appears feasible. These nuclear reactors must be light in weight, have shielding provisions to minimize the radiation exposure of sensitive electronic equipment, instruments, or a crew, and usually require a waste heat radiator, which is vulnerable to meteoroid penetration damage.

The development of the first space nuclear auxiliary power source with a turbine power conversion unit is under way. It uses a 1,200° Centigrade fission reactor cooled by liquid alkali metal, which gives up its heat in a heat exchanger to a closed cycle mercury vapor loop, which powers the turbine. The first unit with 2–3 kilowatt electrical output will be ready soon and will be followed by a more powerful version, producing 30–60 kilowatts. The technical problems are associated with materials, high temperature, radiation, power level control, and long life.

The short duration devices using chemical energy sources such as the fuel cell, batteries, or positive-displacement engines are limited to short duration applications such as a single-orbit satellite or circumlunar navigation; here the fuel weight becomes prohibitive for longer durations. For durations of more than three days a nuclear or solar radiation source is essential.

Research and component developments are under way in all areas of power sources and power conversion; it is too early to rule out any of the methods described.

GUIDANCE

As with vehicles and propulsion, guidance systems and components for space flight have been outgrowths of those used in military ballistic missiles. The principal components, such as gyroscopes, radars, antennae, computers, accelerometers, power supplies, or stable platforms are well understood, and further improvements in their technology are still under way. However, a few new components peculiar to space flight are needed. In space flight, guidance is achieved, in part, far away from the launching station, and for durations which are much longer than those required for ballistic missiles, so that a space vehicle usually has more than one basic guidance system, or the same system is operated in a different way for different parts of its flight.

A guidance system traditionally fulfills three functions: (1) It senses the *actual* vehicle attitude, position, and velocity. (2) It determines the *desired* vehicle attitude, position, and velocity for a given flight mission objective. (3) it determines the necessary corrective maneuvers and issues instruction signals to control the vehicle's attitude, position, and velocity for staying on its intended trajectory, and for achieving the intended mission. Thus a guidance system contains elements which sense (in three dimensions) vehicle attitude, position, or velocity; one

Space Vehicles

or more computers; an accurate clock; and a reference coordinate system. Associated and integrated with the guidance system, but with a separate task and design is the vehicle control system, which physically steers and stabilizes the spacecraft, maintains the proper pitch, yaw, and roll attitude in the presence of external disturbances, and causes the application or termination of forces or torques to the vehicle. Functional elements of both the guidance and control system are often put into one package. Related equipment, such as power supplies, navigational reference devices, ground-support equipment, and checkout systems, is also integrated into the guidance system.

Radio guidance systems can be used to measure range Doppler velocity, angle, and angle rate of change, but are limited by propagation disturbances, signal strength, or line of sight geometry. Inertial systems are completely self-contained and can sense accelerations in all directions, and the attitude of the vehicle, and compute velocity and position, but their errors increase with time. Combinations of radio and inertial guidance seem to be best suited to spacecraft. Novel sensors designed especially for space flight are now being developed and include, horizon scanners (to establish the direction of the vertical in an earth satellite), optical celestial observation of the sun and the stars (for navigational triangulation), magnetic field sensors, and some new types of atmospheric sensors for re-entry.

While the basic inertial and radio guidance systems developed for missiles seem adequate for guidance during the ascent and launch phases, additional guidance corrections and signals are usually required for the midcourse and terminal phases of a space flight, and for attitude control throughout a flight. For a lunar landing, for example, small trajectory corrections are applied during the midcourse flight after the vehicle achieves an earth-moon orbit (that is, after the last launch rocket stage has operated) in order to minimize the errors and thus to select more closely the place of arrival on the moon. This is perhaps best accomplished by radio guidance surveillance of a vehicle equipped with a radio beacon, with computation and command transmission of corrective control information from the ground. During the terminal phase of such a lunar mission map matching and radar surveillance of the moon from close range can be sensors which, together with an inertial system, can be effective in more accurately locating a specific landing spot and triggering the retrorocket. Other typical midcourse and terminal guidance requirements, such as an orbital rendezvous with homing devices, corrections to the orbit of an earth satellite, position keeping of a 24-hour stationary satellite, or angle of attack control on a re-entry vehicle, are needed for different missions.

Attitude control is often required throughout the flight to orient antenna toward the earth for communication, toward the moon for radar reflection; to orient optical equipment and solar cells toward the sun or stars; and to orient cameras, telescopes, or other specific scientific instruments at various periods during a flight. This usually requires some inertial system components; for earth satellite missions it also requires a horizon scanner; for deep space missions a new device, called a celestial rate sensor holds considerable promise. It uses the fact that the drift velocity found by apparent motion of stars, seen through a fixed, vehicle-mounted telescope, is the negative of the angular velocity of the vehicle about an axis normal to the drift direction.

RECOVERY

For those space vehicles which re-enter the atmosphere, it is necessary to develop materials to withstand the high rates of heating and the high temperatures that occur upon re-entry into the atmosphere at high velocity. Similar but different problems exist in some of the engines, in some of the high temperature gas-cooled spaceborne reactors, and in other thermodynamic machinery. The transient heating of atmospheric re-entry bodies is accompanied by ionization of the boundary layer, which in turn causes temporary interference with radio communication. This has led to the development of both metals and ceramics, and combinations of metals and ceramics, which are resistant to high temperatures. The technology of columbium, vanadium, tungsten, and molybdenum base alloys has blossomed with the space age, and has applications in industrial technology. Steel

Space Vehicles

pressure vessels withstanding 200,000 pounds per square inch are another development of the space age.

Several different techniques have been proposed and tried for recovering payloads from satellite orbits or from high velocity ballistic missile trajectories. Some use the principle of a heat sink, which is basically a high density, high conductivity heat sponge, to absorb the energy that is created in the atmosphere boundary layer upon re-entry. Another method allows some of the material in the outer shell of the re-entering body to be melted or sublimed, so that it can absorb additional heat by liquification or vaporization.

This method is called *ablative* cooling. Special glasslike materials are used on the outer surfaces. Although re-entry bodies have been in existence for only a few years, they have already undergone some fairly drastic changes. Today's nose cones are mainly cooled by ablation, instead of the heat-sinks—usually blunt bodies with high drag—used earlier. The modern streamlined aerodynamic shapes permit faster re-entry. Aerodynamic controls on lift-type re-entry body shapes permit maneuvering, and thus selection of the landing area. The heat protection techniques permit the transition of the re-entering body from the outer space regions through the upper layers of the atmosphere. Thereafter some other method of recovering the useful payload must be used, such as a parachute, or aerodynamic control surfaces for achieving a landing, or the capture of the payload by means of aircraft. Successful recoveries of capsules from several Discoverer vehicles and Soviet satellite vehicles have been achieved.

IMPACT ON INDUSTRY

Many of the devices and products specifically developed for the missile and space weapon systems have useful applications in civilian markets and products. For example, compounds of hydrazine, which proved to be a very effective high energy, storable liquid propellant, have pharmaceutical applications. One of the derivatives of hydrazine is used in the poultry industry. A few years ago it was necessary to develop a lightweight, silent, vibration-free solenoid valve to be used as a pilot valve in liquid propellant rocket engines. This same technology has many civilian applications. For example, the heaters used in the average home have new silent solenoid valves which no longer jar the house when the furnace starts, as the old types did.

The space age has brought with it a host of new methodological developments in materials and fabrication processes, which have increasing application in other fields. One of these is the method of roll-forming highly stressed steel cylinders for solid propellant cases without longitudinal welds, which have always been the weak point in the design. Arc plasma deposition of refractories and refractory metals has been useful for the fabrication of coatings of tungsten, carbides, or tantalum to improve or inhibit transfer of heat, wear resistance of surfaces, or to provide special reaction catalysts. Pyrolytic vapor deposition has shown completely unexpected physical properties in materials; for example, graphite deposited from methane on a hot mandrel gives a graphite with tensile strength five to eight times as high as commercial graphite and highly indirectional heat conduction properties; graphite so obtained is desirable for high temperature applications, such as rocket nozzles, furnaces, nuclear reactors, or turbines. Other materials, such as certain oxides, refractory carbides, and boron nitride, can be obtained by this technique of vapor deposition. New manufacturing processes such as electron beam welding (one of the best methods for joining refractory metals), sandwich construction of beam structures, electrolytic etching, and electrolytic deposition of metal, were first developed for missile or space vehicle use, and they are being applied in many other areas.

The micro-miniaturization in electronics certainly was stimulated by the advent of space devices, and this technology is used in such products as portable radio and television sets, office machines, industrial machinery control, and complicated computers. Some of the methods of making lightweight, efficient propellant tanks are used in the tank truck industry, reducing the dead weight of tank trucks, thus saving on operating costs.

Space vehicle systems generally are complicated, technically

Space Vehicles

difficult, and highly accurate. Industry is usually required to build only a few of each major vehicle, payload, subassembly, or component. Complex technology has not only made possible individually more complex space systems, but it has also affected the weapons system technology. The aviation industry has had to shift from an era in which there were relatively simple, mass-produced products to one requiring relatively few diverse articles, but where each one is exceedingly complicated, and requires many skills from different fields of technology and science.

This has changed the structure of the aviation industry; there is a larger percentage of technical and scientific personnel on the staff now, and the emphasis on manufacturing capability and manufacturing space has diminished. The growth of automation was aided by the advent of the space age, and automation is now used industrially for such purposes as making out payrolls, process and production control, and for solving technical and scientific problems. The new emphasis on reliability has introduced a different philosophy of manufacturing. There now is a great emphasis on high-grade quality control redundancy in design, extensive simulation failures, simulation of space environment, statistical planning of test experiments, the development of novel test techniques, and reliability engineering. This basic reliability philosophy has affected many organizations and is being applied to other technologically complex fields.

The government's fiscal year 1962 expenditures for space projects amounted to approximately 2.5 billion dollars, of which 1.8 billion is being spent by the National Aeronautics and Space Administration. This amounts to approximately 3 per cent of the federal budget, and about one-half of 1 per cent of the gross national product. It is estimated that during the next several years the annual expenditures for space projects will rise to around four to five billion dollars, or approximately 1 per cent of our gross national product. This will give employment to about one-half million people and support indirectly several million people, who are required to maintain those directly employed in industry, universities, and government laboratories. It is probable that many times that number will benefit from the commercial and industrial use of products, materials, and ideas started or developed under the research and development programs aimed at space exploration.

One visible impact of the space program on our economy is the transformation that occurred in some of our government organizations and agencies. The complex technology and science involved in space research was one of the primary reasons for the creation of the office of a science adviser to the President, for the statutory establishment of a space council consisting of the principal government people concerned with space policy and headed by the Vice-President, and for the establishment of some two dozen different organizations within the government that are actively and principally concerned with space.

The investigations on missiles and space have been responsible for creating at least two large new communities near the Pacific and Atlantic missile ranges, and space research has augmented directly or indirectly the population in many other centers of our country where factories and laboratories for these devices are, or will be, operating.

The stimulus of space to the growth of research and technology and the molding of our industrial and national economy has probably not been as profound as the impact of space on the individual. The arrival of the space age has widened his horizons, changed some of his thinking, and aroused his interest in worthy new goals, which are truly beyond our own old world.

CHEMICAL AND NUCLEAR ROCKET PROPULSION

RALPH S. COOPER

In the exploration of space, propulsion—the means of getting there—has always been the crucial problem. Inspection of any of our space vehicles shows that the bulk (over 90 per cent) of their volume and weight is devoted to the propulsion systems, mainly to the propellant itself. The extent of space exploration in the future will depend primarily upon the sizes and efficiencies of the propulsion systems which will be developed.

One might wonder whether large-scale space operations are economically feasible in terms of their apparently high energy requirements. But in fact, the electrical energy used by a typical American household during one month is sufficient to put about 75 pounds into orbit about the earth. The same amount of energy is contained in only six gallons (50 pounds) of gasoline plus the oxygen (150 pounds) needed to burn it. Rockets, being far from 100 per cent efficient in transferring energy to the payload, require about five to ten times as much energy as this, but still the requirements are not unreasonable. Furthermore, atomic nuclei represent a very compact, almost limitless, source of energy if we can find ways to utilize them efficiently.

VELOCITY REQUIREMENTS

Space travel is dynamic in the sense that velocities rather than positions are significant. The important effect of propulsion is to change the vehicle velocity, which then results in an appropriate change in position, and thus one usually expresses the propulsion requirements for various missions in terms of a velocity. For example, the velocity required for a low earth orbit is about 26,000 feet per second (or 18,000 miles per hour). In addition, one must lift the vehicle to some height and overcome certain gravitational and atmospheric losses (such as aerodynamic drag). These losses are frequently evaluated in terms of velocity, and included in the mission requirement. Table 1 gives the approxi-

TABLE 1
MISSION VELOCITY REQUIREMENTS

Mission	Velocity* (ft/sec)
Low earth orbit	30,000
Earth escape / Lunar hit	42,000
High earth orbit / Lunar orbit / Mars, Venus probes	45,000
Lunar landing	50,000
Lunar round trip / Escape from solar system	60,000
Mars, Venus round trip	60,000 to 90,000

*Including losses

mate requirements for some missions of interest. A lunar round trip has only twice the velocity requirement of orbital missions, and interplanetary trips need only three times orbital velocity. However, this implies that the rockets must be respectively four and nine times as large as the orbital vehicles for the same payload and propulsion system.

ROCKET PRINCIPLES

Almost all types of rockets are based on the principle of action and reaction and are similar in action to the recoil of a gun or to the motion of a balloon which is rapidly losing its gas. The motion depends upon expelling some material (propellant), be it gas, solid, or charged particles, from the vehicle. Thus rockets carrying their own propellant are able to operate in a vacuum outside the atmosphere, just as a gun's recoil is independent of the air about it. The velocity of the expellant with respect to the vehicle (called the exhaust velocity) is a measure of how effectively the propellant is used, and is comparable to the miles per gallon of an auto engine. The higher the exhaust velocity, the more effectively the propellant is being used, and although this requires more energy per unit mass of propellant, it is advantageous to have high exhaust velocity. Note that the original source of energy does not have to be in the ejected material, although it is for chemically propelled rockets. In our earlier illustrations, the energy was stored in the gunpowder, not the lead projectile, and in the stretched rubber of the balloon, as well as in the compressed gas. The initial gross weight of a rocket for a given payload depends exponentially upon the ratio of the mission velocity requirement and the exhaust velocity, making the results quite sensitive to these quantities. Since the mission velocities are relatively fixed, major reductions of vehicle sizes for given payloads and missions can come only through increasing the exhaust velocity.

The final mass includes the "dead" weight of the rocket (engines, tankage, unused propellant), as well as the payload. For some value of the ratio of velocity requirement to exhaust velocity, the dead weight required for the propulsion system leaves nothing remaining for the payload. This problem is circumvented by jettisoning used portions of the propulsion system, resulting in a number of stages. Usually the tankage and engines of a given stage are dropped when it has exhausted its propellant. Occasionally, as with the Atlas, which drops two of its engines, only portions are released. Staging permits arbitrary mission velocities to be attained, although the payloads may be small. One can find in general an optimum number of stages for a given mission and propulsion system. A high exhaust velocity allows one to use few stages, which results in a simpler, as well as a lighter, vehicle.

CHEMICAL PROPULSION: SOLID PROPELLANTS

Solid propellant rockets are the simplest, and were first historically. They were used in both China and Europe in the thirteenth century. Used sporadically for centuries, they became very popular in warfare about 1800 ("the rockets' red glare") but were displaced by rifled artillery which was much more accurate. They continued in use in a number of minor applications as well as in warfare where much cheap, lightweight, but inaccurate firepower was acceptable. These "powder" rockets con-

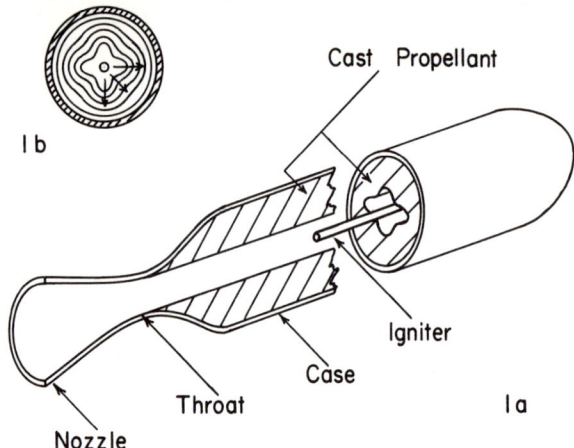

Fig. 1—1a is a typical modern solid propellant rocket shown in cutaway. 1b is a crosssection of a solid propellant rocket showing recession of internal burning surface with time.

tained black powder and, later, smokeless powder grains loosely packed in their cases, necessitating short burning periods to keep the rockets from bursting because of the high temperature.

During World War II, solid rockets using an asphalt base were developed for assisting airplane takeoffs (jet-assisted takeoff or JATO units). This propellant could be cast in a single piece but tended to crack or soften with temperature changes. The development of rubber-based propellants after the war, combined with a design which kept the wall cool, opened the way for large solid rocket engines. The propellant is cast in place in a single mass with a central hole. The igniter in this hole ignites the inside surface, and the burning surface moves outward toward the case which remains cool until the propellant is almost completely burned. During the entire burning period, the case must be able to contain the high pressure (hundreds of pounds per

Chemical and Nuclear Rocket Propulsion

square inch) which originates in the hot gases and is transmitted through the rubberlike propellant. Thus lightweight, high-strength materials are an important requirement for solid rockets, which tend to have high dead weights due to the case. The primary requirement for nozzle materials is ability to withstand extremely high temperatures, and for this purpose, inserts of special refractory materials (for example, graphite or tungsten) are often placed in the nozzle throat.

The propellant itself must contain both a fuel and an oxidizer, which supplies the oxygen for the combustion process. Both can be contained in the same molecule (such as nitroglycerine-nitrocellulose) which forms a homogeneous "double base" propellant. For large motors, a rubberlike fuel with small, discrete particles of oxidizer dispersed throughout ("composite" propellant) is more appropriate, since large pieces can be cast in place with little danger of cracking or softening. The exhaust velocity for such materials is in the range of 7,000 to 8,000 feet per second, which is not as good as many liquid propellant combinations. To place a payload in orbit (at a mission velocity requirement of 30,000 feet per second), four solid propellant stages are required, and such a vehicle (Scout) is being developed by the National Aeronautics and Space Administration. Although it puts less than 1 per cent of its initial weight in orbit, it has many of the favorable characteristics of solids—simplicity, ease of handling and launching, relatively low cost, use of various stages for different missions—which commend it to scientific research work with small payloads (about 100 pounds). The ease of scaling up or clustering solid rocket motors has led to their consideration as large boosters. When used as first stages only, their lower performance and higher dead weight are less significant. In very large sizes, the propellant cost (about one dollar per pound) becomes significant, as does the difficulty of handling the large quantities of potentially explosive material.

LIQUID PROPELLANT ROCKETS

In order to achieve higher exhaust velocities, it is necessary to use propellants with combustion products of low molecular weight, and these chemicals are generally liquid or gaseous at room temperatures. The gases (such as oxygen) would require too much volume and weight to be contained in that state, and thus are liquified and kept at low (cryogenic) temperatures, in contrast to so-called "storable" propellants which are liquid at room temperatures. The first liquid propellant rocket engine was made about 1900, and in the 1920's and 1930's work on them was carried out independently in Germany and in the United States. Little was done in this country, except the work of Professor Robert Goddard.

A variety of applications and propellant combinations were evolved, including an anti-aircraft missile using aniline as the fuel and nitric acid as the oxidizer, aircraft rocket engines utilizing alcohol and concentrated hydrogen peroxide, and finally the V-2, an alcohol and liquid oxygen missile. Based on research of the 1930's, its design was begun in 1938, and the first experi-

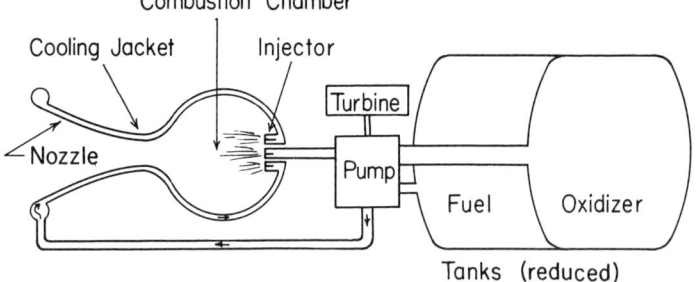

Fig. 2—Simplified schematic of a typical liquid propellant rocket.

mental flight was in 1942. Fortunately, internal political squabbles prevented its completion until late in the war. A 3,000-mile range, two-stage missile was being designed for bombarding the American continent.

With the V-2, rocketry came of age. The Redstone missiles which were used for the United States suborbital manned flights were basically scaled-up versions of the V-2, as were early postwar Russian rockets. Long-range rockets with high explosive warheads are poor, expensive, inaccurate weapons, and probably would not have been developed except for the appearance

Chemical and Nuclear Rocket Propulsion

of nuclear weapons, which gave the final impetus that led rapidly to space exploration capability.

The variety of propellant combinations and types of liquid rocket engines rapidly multiplied, so a brief discussion must select and oversimplify. The propulsion system includes tankage, propellants, pumps to bring propellants to the combustion chamber where they are burned, and a nozzle to expel the gases efficiently. The fluid propellant can be used before combustion to cool the nozzle and combustion chamber, allowing longer periods of operation than uncooled solid rocket motors. The propellants may be forced into the engine under the pressure of gas in the tanks, but since this requires heavier tankage, for large rockets the propellant is pumped into the combustion chamber. Propellants enter through an "injector" which is similar to a showerhead and serves to disperse and mix the propellants for efficient combustion. The pump, which requires considerable power in large engines, is usually powered by a gas turbine. The turbine may have its own gas-generating system or utilize the propellant combustion products to supply its working fluid. The nozzle or entire engine can be swiveled to provide flight control for the vehicle.

The dead-weight of a large liquid rocket propulsion system is 5 to 10 per cent of the stage gross weight. The most commonly used propellant combinations (for example, "RP," a kerosene-like hydrocarbon, and oxygen) yield exhaust velocities of about 10,000 feet per second, while the use of liquid hydrogen as a fuel produces velocities of 14,000 feet per second, which is close to the maximum possible with chemical propulsion. Hydrogen has a very low boiling point and very low density, but its high performance has led to its choice as the fuel for future United States spacecraft. For liquid chemical propulsion, two or three stages are optimum for the earth orbit mission, and only a rocket with high energy fuel and a light structure can place itself in orbit with only a single stage. The liquid propellants are less expensive than solid fuel, but the engines are more complicated and therefore more expensive to develop and build. At this time, it is not clear whether it will be economically feasible to reduce costs by recovering spent boosters for reuse.

OTHER CHEMICAL SYSTEMS

Naturally many proposals have been made for improving the performance of chemical systems by increasing their exhaust velocity, or reducing dead weights or complexity. Specialized systems have been, or will be, developed for particular purposes, including monopropellants (single chemical liquids which decompose to give hot gas), hybrid solid-liquid rockets, and engines with controllable thrust levels for landing. Relatively little improvement can be expected in the exhaust velocity, even with quite exotic propellant combinations, and it is this which primarily determines the performance. Many advances in simplicity, reliability, and structural weight can be expected, but a "breakthrough" in performance of chemical propulsion seems unlikely.

One area where great improvement is possible is in "aerospace" vehicles, which use air-breathing engines (turbojets or ramjets) for a portion of the boost phase of flight. Since only the fuel need be carried in the vehicle, much greater efficiency is possible in the region up to 10,000 feet per second (7,000 miles per hour) which is one-third of orbital velocity. If sufficiently large, high-speed aircraft engines and airframes could be built, they could be used as non-ballistic (flyable), recoverable boosters. In combination with rocket propulsion such "planes" might even be powered into orbit (the so-called "aerospace plane"), although this seems a very formidable task.

ADVANCED TECHNIQUES

Nuclear energy can be a very compact type of almost limitless energy, and it is natural to seek some way of utilizing it for space propulsion. Any form of rocket will require some form of propellant to provide the thrust by being expelled from the vehicle, but with a separate energy supply, this could be used much more effectively (that is, with higher exhaust velocity). There are many methods of nuclear propulsion, with efficiency and complexity generally increasing together. Emphasis here is on those which are closest to becoming a practical reality.

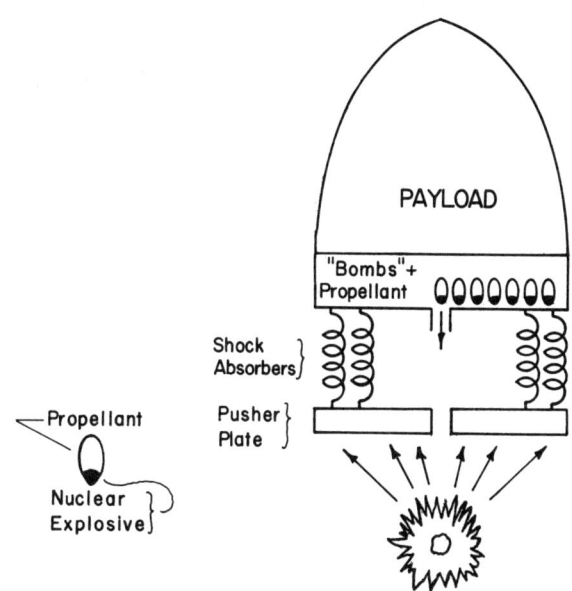

Fig. 3—A nuclear explosive-propelled vehicle. The nuclear explosive heats the propellant, which impinges on the "pusher plate," transferring momentum to the vehicle. A set of shock absorbers smooths out the force on the payload.

The simplest and most straightforward way of using nuclear power in a rocket is to replace the liquid rocket combustion chamber with a nuclear reactor to supply heat to the propellant. The reactor is an array of solid nuclear fuel elements containing a fissionable fuel. When the reactor is brought to power, the heat generated in its fuel is transferred directly to the liquid propellant which is pumped through the reactor. The liquid is vaporized, heated to a very high temperature, and expelled through a nozzle to provide the thrust. Since the heat energy is supplied by a source independent of the propellant (rather than by the propellant's chemical energy), one has a freer choice of propellant.

By choosing hydrogen, which has the lowest molecular weight, one can readily achieve exhaust velocities of 25,000 to 30,000 feet per second, about twice those of the best chemical propulsion system. This high performance is partially offset by the heavier dead weight (10 to 15 per cent necessitated by the reactor and hydrogen tankage) and by the complications arising from the nuclear radiation emanating from the reactor. Nevertheless, the high performance is invaluable for missions in the interplanetary class and useful for less ambitious ones.

The nuclear reactor is basically a simple device, and its chain reaction can be easily controlled through the movement of neutron-absorbing materials in the core (the fuel-bearing region) or the reflector (an outer layer of material which helps to keep the neutrons from escaping). Since there is no combustion, explosions are unlikely, and the nuclear engine should prove to be quite reliable from that standpoint. To obtain the high thrust and high exhaust velocity, the reactor must run at much higher power and temperature than do ordinary power reactors, but this is partially ameliorated by the short lifetime (about 10 minutes) required of rocket reactors. Each cubic foot of the reactor core must generate energy equivalent to the electricity used in many thousands of homes. Furthermore, it must do this while much of the reactor core, several cubic feet and several thousand pounds, is at the temperature of an electric light bulb filament (2,000° to 3,000° C.)! This is far above the melting points of most common materials (such as steel, quartz sand, and most refractory materials) and limits the choice of fuel elements. Graphite (the most familiar), tungsten (used in lamp filaments), and a few metal carbides are about the only candidates. These must contain the fissionable material in a refractory form such as uranium oxide or carbide, which also have high melting points.

The liquid hydrogen presents several difficulties. First, it is only one-fourteenth as dense as water (or most chemical fuels) and thus requires relatively larger and heavier tankage. Second, it boils at $-423°$ F, which is close to absolute zero ($-459°$ F.), and thus requires special insulating and handling techniques. It is cold enough not only to liquify air, but to freeze it solid. Nevertheless, hydrogen is being used as a liquid fuel in the chemical rocket program, and techniques are being developed which should make its use routine.

The nuclear radiation presents a number of problems, but these can be met with straightforward solutions in most cases. It has been shown that even with regular launchings in the atmosphere, the world-wide contamination would be negligible. Hazards for manned operations must be minimized, but the

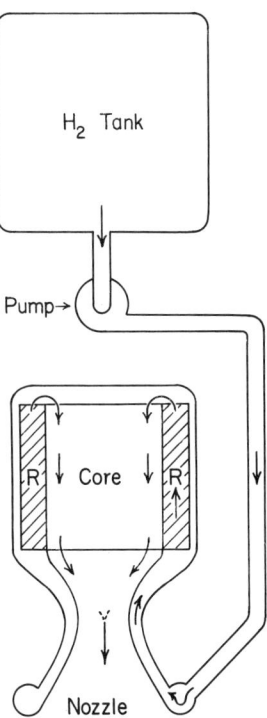

FIG. 4—A simplified schematic diagram of a hydrogen-fueled nuclear reactor propulsion system. The hydrogen is pumped from the tank up through the nozzle and reactor reflector (R), cooling them and being vaporized; it then flows down through the reactor core and is there heated to a high temperature and expelled from the nozzle.

problems may not be much different from those associated with chemical propulsion. Manned space flight may require extensive shielding against space radiation, which will be effective against the nuclear engine as well.

The exhaust velocity for nuclear propulsion is about equal to the velocity increment needed to achieve earth orbit, and therefore a single nuclear stage is capable of going into orbit with considerable (about 20 per cent) payload. For difficult missions, only about half as many nuclear stages as chemical stages need be used, increasing reliability and decreasing launch costs. Finally, operation in space reduces many of the radiation problems and weight penalties associated with nuclear propulsion (lower thrust, lower weight engines can be used, lighter structured hydrogen tanks employed).

ADVANCED NUCLEAR PROPULSION

Since the nuclear heat exchanger engine uses less than 0.1 per cent of the fission energy available, we can see that only the beginnings of nuclear propulsion have been touched upon. The problem lies not in obtaining the energy as much as in dealing with the higher temperatures involved when this energy is transferred to the propellant. An additional incentive for seeking temperatures above 3,000° C. is the dissociation of the hydrogen molecules into hydrogen atoms; dissociation occurs over a range of temperatures, which allows much greater storage of energy in the propellant at these temperatures. This could lead to exhaust velocities of up to 50,000 feet per second. There has been hope of making gaseous core reactors, operating at up to 10,000° C., but the problem of separating the gaseous fuel from the propellant appears insurmountable if reasonable thrust is desired. One possibility for circumventing the material temperature problem is the use of small nuclear explosions in what might be called an "external combustion engine." The nuclear explosive heats the propellant behind the vehicle. The propellant impinges on a large, heavy "pusher plate," transferring momentum to the plate which is coupled to the rest of the vehicle through a shock-absorbing system. Space ships utilizing such a propulsion system could carry millions of pounds of payload throughout the solar system with quite small nuclear explosives.

There are no workable methods of utilizing nuclear fusion (thermonuclear) energy for space propulsion presently in sight.

PERFORMANCE COMPARISON

There is as much variety in space propulsion as in surface transportation, and consequently, there are varied performance levels and areas of application. The method of propulsion used is dependent upon availability, cost, or criteria other than performance, simply on a gross weight basis. Other aspects will be discussed qualitatively. Table 2 gives approximate representative parameters for four propulsion systems currently under development. The gross weights of vehicles to send a 10,000 pound payload on various missions are given in Table 3. A

TABLE 2
HIGH THRUST PROPULSION SYSTEM PARAMETERS

	SOLID	LIQUID $RP\text{-}O_2$	HIGH ENERGY LIQUID $H_2\text{-}O_2$	NUCLEAR H_2
Exhaust velocity, ft/sec	8,000	10,000	13,500	27,000
Per cent dead weight	8	5	7	15

TABLE 3
PERFORMANCE OF HIGH THRUST SYSTEMS

Vehicle Gross Weight for a 10,000 Pound Payload

MISSION	ΔV, ft/sec	SOLID	$O_2\text{-}RP$	$H_2\text{-}O_2$	NUCLEAR H_2
Earth orbit	30,000	780,000	280,000	130,000	43,000
Lunar round trip	60,000	60,000,000	8,000,000	1,700,000	190,000
Martian round trip	90,000	..	..	22,000,000	800,000

10,000 pound payload was chosen for consistency and as a reasonable manned vehicle size, although the more difficult missions may require greater payload for radiation shielding and life-support systems. Results are scalable to other payloads. The necessity for staging leads logically to the use of early, low-power versions of advanced propulsion systems in upper stages, resulting in hybrid vehicles with intermediate performance. Each replacement of a chemical by a nuclear stage leads to reductions of two and one-half in vehicle gross weight as can be seen in Table 4 for the lunar mission. Thus the booster size can be

TABLE 4
Vehicles for a Lunar Mission
55,000 Pounds Landed on the Lunar Surface

Vehicle	All Chemical 4 Stages	3 Chemical 1 Nuclear (3d Stage)	1 Chemical 2 Nuclear Upper Stages	All Nuclear 2 Stages
Vehicle gross wt/(lbs)	10,000,000	4,000,000	1,700,000	860,000
Vehicle dead wt.	500,000	220,000	110,000	125,000
Propellant volume, ft^3	200,000	100,000	80,000	160,000

reduced and the payload increased considerably with nuclear upper stages. For manned vehicles, a chemical last stage would be desirable to act as an escape vehicle, and for the shielding its propellant would provide.

There are no unclassified performance figures available for the nuclear explosion scheme, and low-thrust, nuclear-electric propulsion performance depends crucially on the specific weight of the power plant. If the desired values can be achieved, electric spacecraft which start from orbit could carry about one-third of their gross weight as payload on interplanetary round trips.

To summarize the situation, most of these propulsion systems can be used exclusively for any vehicle, but combinations will be used which reflect their attributes and state of the art. Liquid propellant rockets will be the most used of the 1960's, with high-energy propellant being used in upper stages. Solid propellant rockets will be used for small final stages (for example, retro-rockets), low total cost, low payload research rockets, and possibly as large, first-stage boosters. Nuclear propulsion will be used in upper stages for difficult missions (lunar and interplanetary). The low-thrust electric propulsion systems will be limited to interplanetary missions which start from orbit.

AUXILIARY POWER SUPPLIES

Up to the present, electrical power has been supplied to spacecraft mainly by chemical batteries or solar cells. The former are relatively heavy per unit of output, and have short lifetimes. The latter are limited to low powers, which will fluctuate with the spacecraft's orientation and position, and are affected by radia-

Chemical and Nuclear Rocket Propulsion

tion. Some of these difficulties are relieved by using the two in conjunction, allowing the solar cells periodically to charge the batteries, which supply continuous power. Nuclear energy represents a way to circumvent the lifetime and power limitations. One method, which has already been put into practice, is to use the heat generated by radioactive isotopes to supply energy to thermoelectric generators. These convert heat into electricity in the same manner as do temperature-measuring thermocouples. Radioisotope sources are somewhat limited in power (several hundred watts) but can have lifetimes ranging from 100 days to 100 years or more in practice, depending upon the isotope chosen.

For high powers (kilowatts) and long times (years) nuclear reactors are the only practical source. A nuclear electric power supply must include power conversion equipment, radiators to reject unusable heat, and possibly some shielding, as well as the reactor. This leads to system weights of the order of 1,000 pounds, useful only in large payloads. At present, only rotating electric generators are sufficiently developed to handle the high power. These will have a metal vapor (such as mercury) heated by the reactor, to power a turbine which drives a generator. The vapor is condensed and cooled in the radiator to complete the cycle. Eventually, high-power, thermoelectric conversion systems will be developed with lower weight, higher efficiency, fewer moving parts, and greater reliability than the turbogenerator system.

STATE OF THE ART AND PROSPECTS

In 1961, most large space boosters used liquid propellants with exhaust velocities of less than 10,000 feet per second. Our largest, the Atlas-Agena, can place over 5,000 pounds in orbit, compared to 14,000 pounds for the Russian vehicle, which is therefore probably a similar system of two or three times the size of the Atlas. A high energy liquid (hydrogen-oxygen) propellant upper stage (Centaur) is being developed for the Atlas and will be test flown in 1962. It will double the orbital capability and allow sending appreciable payloads to the moon for the Ranger and Surveyor instrumental lunar exploration programs. Table 5 presents information for vehicles in use or being developed by the National Aeronautics & Space Administration.

TABLE 5
NASA SPACE VEHICLES

Vehicle	Payload–lbs Orbit	Escape	Type
Scout	150	....	Four solid propellant stages
Thor-Agena B	1,600	....	Two liquid propellant stages
Atlas-Agena B	5,000	750	Two liquid propellant stages
Atlas-Centaur	8,500	2,500	Centaur uses H_2 fuel
Saturn C-1	20,000	5,000	Interim vehicle
Saturn C-5	200,000	80,000	Apollo booster

Larger hydrogen-oxygen engines and upper stages are planned for the 1960's. The Saturn C-1, our largest booster under construction, uses the standard liquid propellants and has a combined thrust of 1.3 million pounds. It will be tested with high-energy upper stages in 1963, and can place 20,000 pounds in orbit. It will be used for the Apollo manned space vehicle program. Our largest engine under development is the 1.5 million pound thrust F-1, which will be clustered to produce a booster of 6 to 12 million pounds for extensive space operations. The upper stages of the configuration which is now being considered for future missions will be high-energy chemical or nuclear. With experience gained in the military rocket programs, the solid propellant groups have been scaling up their engines (now up to 500,000 pounds thrust) in the hope of producing a very large clustered solid booster in a much shorter time than required for the development of the more complex liquid systems. Thus the advances in the chemical propulsion field will be principally in switching to the high-energy propellants (hydrogen-oxygen) and making larger sizes of the present vehicles. Nuclear propulsion is in the early stages of engine development. Three reactor tests (Kiwi A series) were held in 1959–60 with promising results. These used graphite fuel elements loaded with enriched uranium, and gaseous hydrogen as the propellant. During 1961–62

Chemical and Nuclear Rocket Propulsion

a new series (Kiwi B) of tests will be held, with the design of a flight test engine as an objective. Liquid hydrogen will be used as well as components (such as the nozzle) more suitable to an actual rocket engine. Considerable study has gone into flight testing, which will probably involve a nuclear upper stage to be tested and used in the mid-1960's.

Development of various engines and power supplies for electrical propulsion is under way, with plans to test the engines for short flight periods in space in 1962 and 1963, using the Scout solid rocket as a booster and chemical batteries for the power supply. More extensive tests and use will come in the mid-1960's and depend primarily on the nuclear electric power sources. Nuclear explosion propulsion is at an early stage of research, but the scheme has been checked with a 3-foot diameter, 300-pound scale model, using 3-pound high-explosive charges.

We can expect liquid fuel chemical rockets to be the workhorses in the 1960's, with nuclear upper stages for many missions, and possibly solid propellant first stages. Electric propulsion may be used, starting from orbit, for deep space missions. Some of the more advanced schemes, such as nuclear explosion propulsion, may come to fruition near the end of the decade.

DEEP SPACE PROPULSION SYSTEMS

A. THEODORE FORRESTER

Chemical rockets are not optimal engines for many space missions, especially interplanetary explorations, since the required total weight of propellant is enormously greater than the weight of payload which can be carried. Also, for the more distant planets, the requisite mission times of decades or more become almost too long to be considered feasible.

It is natural to ask whether there exist, at least in principle, propulsion systems which offer the hope of eliminating these difficulties. Basically the question is: Is it possible to get more thrust per unit of mass exhausted than one gets from chemical rockets? Or even further: Is it possible to produce a thrust in space without expenditure of shipboard mass at all? In these terms the answer is unequivocally yes. The impulse per unit mass obtained from a rocket exhaust is simply the average exhaust velocity, and it is perfectly clear that it is possible to produce particles whose velocities are greater than those which exist in flames. Even the second part of the question raised has a positive answer. It is possible to produce a thrust in space without exhausting part of the ship's mass at all, namely, by the process of solar sailing. The nature of the question which must be raised becomes then very different. It is necessary to ask: Is it possible to obtain enough thrust with any of the methods envisioned for a vehicle of given size to produce accelerations in space which make it possible to accomplish missions in reasonable times and with reasonable payloads? Reasonableness will be judged, of course, by comparison with what can be obtained with chemical rockets.

One solution lies in the use of gases of low molecular weight heated in a nuclear reactor as described in the chapter by R. S.

Deep Space Propulsion Systems

Cooper (p. 319, above), but nuclear-thermal propulsion systems do not provide the ultimate answer. For deep space missions the mass of propellant utilized will still be vastly greater than the deliverable payload, and even if we choose to ignore the formidable difficulties which remain in the development of a practical nuclear rocket, it is necessary to consider techniques of obtaining exhaust velocities higher than those obtainable by exhausting very high temperature hydrogen from a nuclear reactor. These potential deep space propulsion systems are either electrical rockets in which electrical power generated on the vehicle is used to accelerate charged particles to the desired velocities; photon rockets in which energy generated on the vehicle is used to create a beam of electromagnetic energy (little hope of success); or photon "sailing" in which the beamed energy in space is redirected to produce the desired thrust. All these are inherently low thrust systems and can be used only as the final stage of a multistage booster system. They will never be useful for lifting a vehicle unaided from rest on the surface of the earth.

REQUIRED PARAMETERS

It does not lie within the scope of this chapter to perform thorough analyses on the systems under consideration here, but certain conceptual and quantitative considerations about the parameters which are involved must be indicated. First, it must be recognized that there is a minimum acceptable acceleration if the trip times are not to become excessive. In Table 1 the various mission times are shown in comparison with mission times which can be obtained using only chemical stages. Table 1 shows that accelerations in the range 1×10^{-4} to 3×10^{-4} g. produce remarkable decreases in many mission times and even accelerations as low as 5×10^{-5} g. may be useful, but accelerations much below 5×10^{-5} g. result in mission times so long as to render such systems impractical. Mission times alone are not, of course, the only considerations. For many types of missions it would be appropriate to lengthen mission times, if by so doing one could improve the payload capabilities of the vehicle. One of the great advantages of electrical propulsion

TABLE 1*
SPACE MISSIONS

	LOW THRUST SYSTEMS		CHEMICAL ROCKETS
Mission	Acceleration (g)	Time (days)	Time (days)‡
Venus Capture†	2×10^{-5}	700	139
	3×10^{-5}	500	
	5×10^{-5}	325	
	1×10^{-4}	130	
Mars Capture	2×10^{-5}	800	260
	3×10^{-5}	600	
Mars Orbiter§	5×10^{-5}	410	
	1×10^{-4}	240	
Jupiter Flyby	2×10^{-5}	1,300	1,060
	3×10^{-5}	1,000	
Jupiter Capture	5×10^{-5}	900	
	1×10^{-4}	575	
	2×10^{-4}	375	
	3×10^{-4}	300	
Pluto Flyby	1×10^{-4}	1,250	43 years

*Mission times obtained from E. Speiser, California Institute of Technology Jet Propulsion Laboratory.
†Capture refers to a very eccentric orbit barely captured by the planet.
‡For minimum energy transfer.
§Orbiter refers to a low level circular orbit.

systems is this ability to carry heavier payloads if longer time is allowed.

With simplifying assumptions, the acceleration can very easily be related to other parameters in which we are interested, such as exhaust velocity, mission time, and power supply weights. The acceleration as given by Newton's law $a = F/M$ leads us to an average acceleration,

$$a = \frac{qv}{\alpha (\tfrac{1}{2} qv^2) + \tfrac{1}{2} qT}, \qquad (1)$$

where q is the rate at which mass is expelled at an average exhaust velocity v, α is the power supply mass per unit power

Deep Space Propulsion Systems

consumed, and T is the total thrust time. In setting up this expression it is assumed that the total power requirement is equal to, or at least proportional to, the exhaust power requirement, $\frac{1}{2} qv^2$, and that the spread in exhaust velocities is low enough that it is not necessary to be concerned about distinctions between average and root mean square velocities. The $\frac{1}{2}$ term in the denominator represents propellant mass as an average between the initial propellant mass qT and the final value 0.

If the expression is rewritten as

$$a = \frac{2}{\alpha v + \dfrac{T}{v}}, \qquad (2)$$

it becomes immediately clear that for a given mission time there is a maximum average acceleration. If v is very large the power supply weight dominates and obtainable acceleration decreases. If v is small (as for chemical rockets), the propellant weight dominates and acceleration decreases. The optimum occurs when the two terms in the denominator are equal; this corresponds to a vehicle which is two-thirds propellant and one-third power supply at takeoff. Other interesting facts emerge. For example,

$$a_{\max} = \frac{v_{\text{opt}}}{T}, \qquad (3)$$

suggesting an exhaust velocity which is the same as the velocity increment given the vehicle. While this result is not strictly correct, based as it is on an approximate treatment without consideration of payload weights, distance traversed, effect of gravitational fields, or numerous effects which may be important, it does give an insight into certain aspects of the optimization problem. More precise treatments will also show that the desired exhaust velocity is approximately the incremental velocity which must be given the vehicle to execute the mission. Even for the shortest deep space missions, such as one-way probes to our nearest planetary neighbors, this incremental velocity must be of the order of 30,000 feet per second or 10^6 centimeters per second, and becomes much larger when consideration is given to problems of going to the distant planets, or of going into orbit around another planet and then returning to earth. In general it may be said that the desired range of exhaust velocities for electric propulsion are

$$2 \times 10^6 < v < \times 10^7 \text{ cm/sec} \tag{4}$$

or in terms of specific impulse (I_{sp})

$$2000 < I_{sp} < 20{,}000 \text{ sec.} \tag{5}$$

These are large velocities compared to thermal velocities obtainable at reasonable temperatures but are quite modest from other considerations. For example, they correspond to protons with an energy of 2 to 200 electron volts, or uranium ions with an energy of 500 to 50,000 electron volts, and the electric rocket appears feasible in principle, at least, if power supply weights can be made small enough to permit adequate acceleration.

From equations (2) and (3) we see that the maximum acceleration obtainable is given by

$$a_{\max} = \frac{1}{\alpha \ (\Delta V)} \tag{6}$$

where $\Delta V = aT$. If we require $a \geq 10^{-4}$ g. and $\Delta V \geq 10^6$ cm/sec, we are led to a requirement on the power supply of $\alpha \leq 200$ lb/kw.

Actually more precise considerations lead to lower values. Probably $\alpha \leq 50$ lb/kw is more realistic if useful payloads are to be delivered.

An especially important parameter which has not been explicitly considered is the payload fraction. The payload of an electric propulsion system can always be increased by increasing the mission time, a fact not true for chemical rockets, where for a given specific impulse, there exists a maximum payload for any given mission. The limitation on the payloads is fairly generous and it can be demonstrated that for the space power supply

Deep Space Propulsion Systems 341

weights which can be anticipated by the late 1960's a very significant fraction (25 to 75 per cent or higher) of the weight boosted into a low-level orbit will be available as payload weight for deep space missions.

POWER SOURCES

Since the constant α is such an important parameter it is worthwhile to compare the various power systems to see if any can satisfy the demand made upon the power supply for a space propulsion system. Batteries are inadequate. The most advanced form of battery contains 50 to 100 watt hours per pound, and if we take 200 pounds per kilowatt as a maximum value of α, we can get at most about 20 hours of running time out of batteries. Neither can chemical energy be stored in any form. Chemical reactions release from 1,000 to 2,000 watt hours per pound, enough at most for 100 hours of operation, completely inadequate for any deep space missions at accelerations of 10^{-4} g.

Only two types of primary sources of power appear suitable for space propulsion—solar radiation and nuclear power. The specific power values for solar electric power supplies depend upon efficiency of the thermal-to-electric conversion techniques, the mass per unit area of the radiation collector, and a number of other factors. At lower power levels solar power sources appear to have outputs comparable with those of nuclear-to-electric power supplies, but at the high power levels the nuclear sources of the future appear much better for deep space propulsion systems. In the vicinity of the earth, where the incident radiation density is approximately 1.4 kilowatts per square meter, the values of α which may be anticipated for various thermal to electrical conversion techniques are summed up in Figure 1. By the year 1970 it is possible that solar power supplies in the power level up to 10 or 100 kilowatts may be light enough in weight to be useful for propulsion.

Although it seems that solar electric power sources may eventually be feasible for electric propulsion systems, by far the most promising sources for this purpose are those which use a nuclear fission reactor as a heat source. Several are due to be

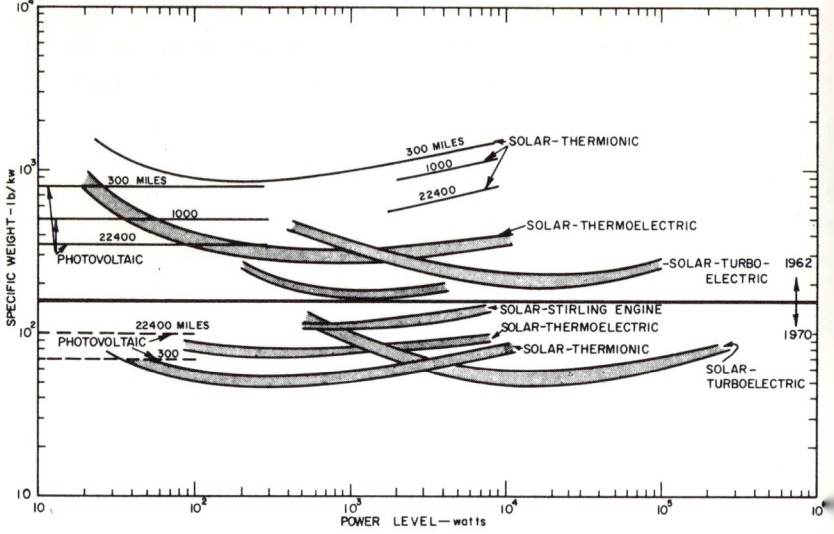

FIG. 1—Power supply mass per unit power consumed (in pounds per kilowatt) as anticipated for several solar thermal-to-electric power conversion systems.

tested soon, and to be operational at electrical output power levels of 3 kilowatts in 1964, 30 kilowatts in 1965, and higher levels soon after. At the low power levels the power supply weights will be about 200 pounds per kilowatt, in the 30-kilowatt range about 40 pounds per kilowatt and will decrease considerably at higher power levels.

Except for a small demonstration system (Snap X, 300 watts, thermoelectric power conversion) the other systems (Snap II at 3 kilowatts, Snap VIII at 30 kilowatts extendable to 60 kilowatts) utilize a mercury vapor turbine and rotating electrical machinery to convert the thermal power to electrical power. The use of thermionic energy conversion for such systems offers great hope that in the future rotating equipment may be eliminated and a higher temperature cycle may be used. Higher temperatures in space power systems are extremely important, not so much because of the greater Carnot efficiencies which are made possible, but because a higher temperature radiator makes it possible

Deep Space Propulsion Systems

to get rid of heat at the low temperature end of the cycle from smaller radiators. Since the radiator is a major component of these systems this is an important advantage. At present thermionic converters are either too inefficient or too short-lived but offer great hope for the future.

ION MOTORS

The term "ion motor" (or engine) usually refers to engines in which positive ions are electrostatically accelerated to the desired range of exhaust velocity. A schematic representation of such an engine is shown in Figure 2. In this engine ions are first

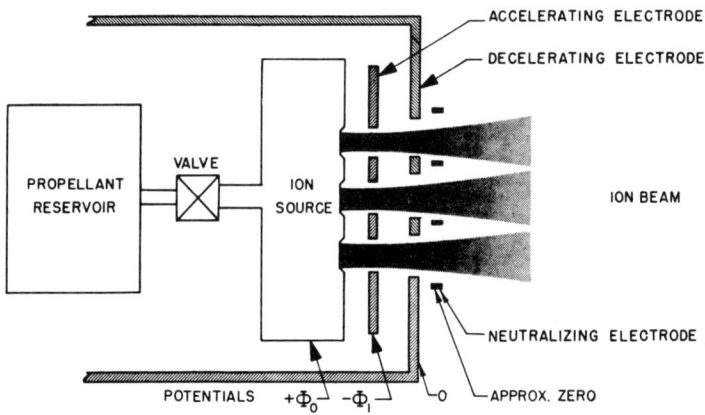

FIG. 2—Schematic diagram of an ion engine.

accelerated through a potential difference $\Phi_0 + \Phi_1$ then decelerated by an amount Φ_1 to a final energy $e\Phi_0$. Electrons from the thermionically emitting neutralizing electrode are drawn into the positive ion beam to provide a neutral plasma as the exhaust.

The use of accelerate-decelerate geometry is required to prevent the injected electrons from flowing "upstream" to the ion source, with consequent production of thermal damage. It serves also to provide a higher voltage across the accelerating gap than is obtainable with Φ_0 alone, Φ_0 being determined by the desired specific impulse and the mass of the ion used.

Before discussing the stage in development of such engines, one should examine the general question of the ionic mass desired for this type of engine. It has been shown that the desired range of exhaust velocities is approximately 2×10^6 to 2×10^7 centimeters per second. The corresponding accelerating voltages which are required are shown as a function of atomic weight in Figure 3. The region in the lower left has been

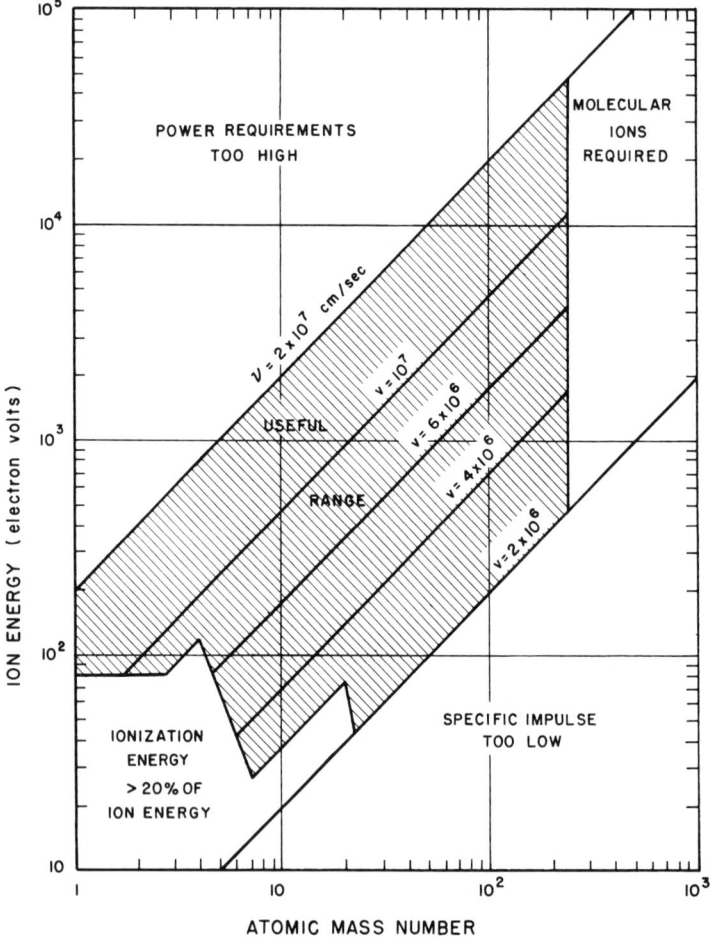

Fig. 3—Accelerating voltages (ion energy in electron volts) plotted against atomic mass number.

excluded from the useful area on the basis of the energy required to generate the ions. Even in good ion sources the energy used to create the ion is about 50 times the ionization energy rather than the 5 times used in this curve, and it is probable that an even larger region should be excluded. This point, however, need not be belabored since there are other reasons discussed below for considering only heavy ions for ion motors.

It is important to realize that the forces on an ion engine are electrostatic; that is, one may say that the impulse carried off per second by the beam must be the integral over the electrode surfaces of the electric stress given in meter-kilogram-second units by:

$$\frac{F}{A} = \frac{1}{2} \varepsilon_0 E^2$$

If we take 10^5 volts per centimeter as a practical upper limit to E we are led to an upper limit to the thrust density of about 10 pounds per square foot.

Considering a space ship as a 10^{-4} g. vehicle, a 10-pound thrustor is the engine for a 100,000-pound vehicle. Clearly the limiting force density is more than large enough. Thrust densities 1 per cent as great are probably satisfactory, but even this corresponds to fields of 10^4 volts per centimeter. To obtain such fields requires either high voltages or very small spacings. If one would like dimensions which are of the order of ½ centimeter or greater, then it is necessary to use potentials of 5,000 volts or more, for which clearly only the very heaviest atomic ions are appropriate. A source of molecular ions or of colloidal particles is desirable but such sources generally yield, in addition to the low specific charge particles sought, many atomic ions which ruin the engine efficiency.

Another reason for the use of heavy ions lies in the neutralization problem. If the emerging ions are not to be turned around by large potentials within the beam, it is necessary that the beam be neutralized on a microscopic scale by negative charges, presumably electrons. There have been questions about the possibility of neutralizing a beam in space, and it can be shown that the difficulty varies inversely as the square of the mass of the ion; for example, if the ion mass is doubled it is necessary to go to four times the thrust at a given specific impulse to duplicate the space charge density. In the author's opinion the reasons for believing that the achievement of neutralization is a trivial problem outweigh the reasons for concern, and for purposes of this book do not warrant further discussion.

The reasons for using high voltage heavy ions are quite compelling but they cause the disadvantage of the erosion of electrodes because of serious sputtering produced by these ions. The problem depends a great deal on electrode material and other factors but, to give a rough example, to attain lifetimes of the order of one year it will be necessary to get ion beam interceptions which are of the order of 1 ion in 10^5 beam ions. The focusing problem has not been extremely difficult, but charge transfer interactions between beam ions and the small amount of neutral vapor in the acceleration gap lead to slow ions generated throughout the gap. These inevitably strike the accelerating electrode and result in erosion which may limit the lifetime of ion engines. Engines operating now appear to have continuous operating lives of approximately 1,000 hours but the limitations in present laboratory tests are not as basic as the limitation because of charge transfer. If, as development proceeds, lifetimes of electrodes do not become as long as needed, procedures for the occasional changing of electrodes may have to be used.

The ions may come from either a surface ionization source or a gas discharge source. In the former case an alkali metal vapor —cesium, the heaviest and most easily ionizable, is the obvious choice for propulsion—is diffused through a heated, porous tungsten diaphragm from which it emerges as ions, which are then accelerated to provide the engine exhaust. The most advanced ion engine to date is operated at Electro-Optical Systems, Inc. Its porous tungsten ionizer is a close-packed array of $3/16$-inch diameter discs, and the accelerating and decelerating electrodes are parallel plates with colinear holes.

Such an engine produced a thrust of 3.2 millipounds at a power efficiency (beam power divided by total delivered power) of about 65 per cent with about 8,500 seconds of specific impulse, and a propellant utilization efficiency of about 99 per cent in January, 1961.

The most advanced gas discharge ion engines are the engines of Kaufman and associates at the National Aeronautics and Space Administration's Lewis Laboratories, Cleveland, Ohio. In this ion source shown schematically in Figure 4) electrons are

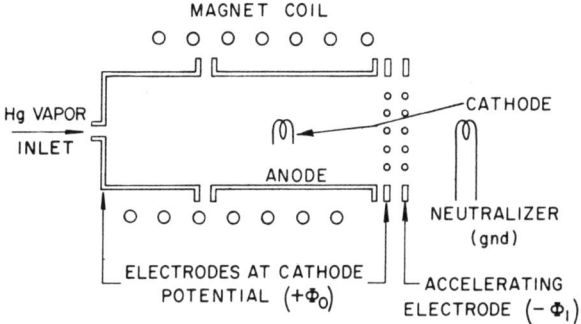

FIG. 4—Diagram of a gas discharge ion engine.

made to oscillate between electrodes at cathode potential while working their way across a magnetic field to the anode. Those mercury ions formed in the discharge which drift to the holes in the screen facing the accelerating electrode are accelerated to produce the ion exhaust. Thrust levels of 20 millipounds at power efficiencies of 80 per cent and specific impulse values of 8,300 seconds have been reported. Propellant utilization efficiencies are about 80 per cent. For gas discharge ion sources it is necessary to be concerned about the increase of specific charge brought about by multiple ionization, but in the Lewis Laboratory work, only negligible amounts of doubly charged ions have been found (20 per cent under some conditions, but generally much less). Even if corrections in the reported efficiency are made for the 20 per cent neutrals and very small percentage of positive ions, the power efficiency remains higher than for the surface ionization sources.

When a detailed comparison between these sources is made on the basis of projected power efficiency, propellant utilization efficiency, ease of construction, ease of control, anticipated lifetime, and other features, each can be demonstrated to have relative advantages, and it does not appear possible at this time to make a clear-cut decision between them. Possibly each will have a domain of operation in which it will be superior.

ELECTROTHERMAL ENGINES

A second method of utilizing electrical power to produce a rocket exhaust is by heating a lightweight gas. Heating may be accomplished by passing the gas over refractory resistance elements or by heating the gas in an electric arc. In the latter case higher specific impulse values may be reached, since the gas may be hotter than the solids in which it is confined, but the efficiencies achievable with either of these techniques at high specific impulse are severely limited by the disassociation and ionization which occurs. It can be shown that high efficiencies can hardly be expected for specific impulse values above about 1,200 seconds, and these are not appropriate for the deep space propulsion systems considered here. Neither do they appear very interesting as primary propulsion systems in comparison with nuclear-thermal propulsion schemes. Their domain of application appears limited to near space missions requiring very fine thrust control.

MAGNETOHYDRODYNAMIC THRUSTORS

In this category there are many possible types of engines, a few of which are shown schematically in Figures 5 through 8. In the crossed field engines (Fig. 5) the plasma is accelerated by $j \times B$ forces which accelerate the ions, and by their coupling the gas to the un-ionized portion of the exhaust of the plasma generator. Some very promising results have been obtained with devices of this type and a thrust of 3.6 pounds at a specific impulse of 1,600 seconds and a power efficiency of 54 per cent has been reported by S. T. Demetriades of the Northrop Corporation. Lifetime of these devices is a problem, but continuous operation has been sustained for several hours. The specific impulse is low, but it is anticipated that the specific impulse values can be pushed to much higher values.

In the pulsed coaxial guns (Fig. 6) a breakdown is initiated between two cylinders, and the resultant plasma is accelerated

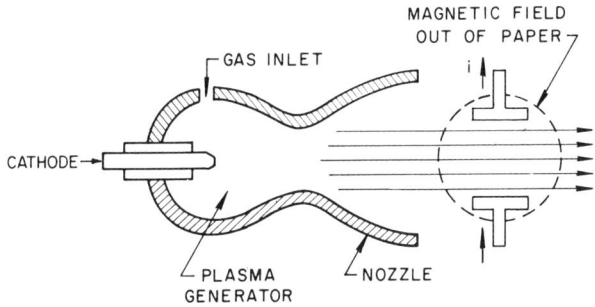

FIG. 5—Diagram of a crossed magnetic field engine.

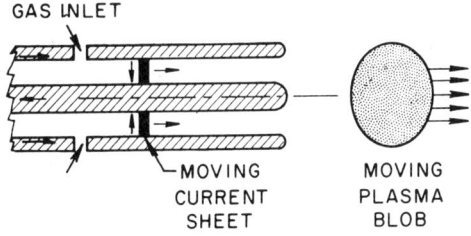

FIG. 6—Pulsed coaxial plasma guns.

toward the end of the cylinder away from the source of power by the self-magnetic field, sweeping out with it all the gas introduced into the tube before the breakdown. In the Republic Aviation configuration (shown in Fig. 7) the plasma is generated as a cylindrical sheet. Because the self-magnetic field in this case forces the plasma inward before it is hurled out through the nozzle, this configuration has been called the "pinch" engine.

A thrust of 0.01 pound with a power input of 1 kilowatt specific impulse of 2,000 seconds, and power efficiency of 50 per cent, has been reported for the Republic "pinch" engine

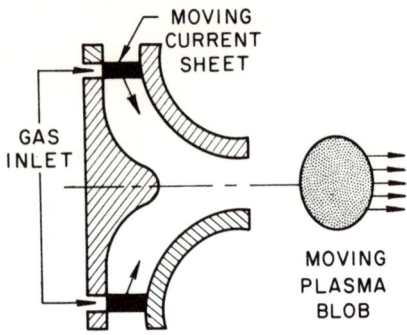

Fig. 7—"Pinch" plasma engine.

(Fig. 7). The specific impulse is somewhat low for deep space propulsion, but there is no fundamental reason why this type of engine cannot produce much higher specific impulse values. Values as high as 15,000 seconds have been reported for coaxial guns, and there appears no reason why these cannot be made as efficient as the pinch guns, or why the pinch guns cannot yield comparable specific impulses.

The engine shown in Figure 8 is sometimes called the peristalsis, or traveling wave, accelerator. In this type of engine a vary-

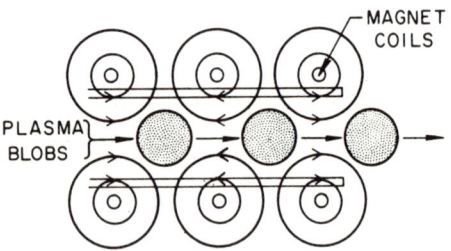

Fig. 8—Peristalsis (traveling wave) accelerator.

ing magnetic field is made to move along a tube either by successive excitation of coils around a tube, or by the proper phasing of radio frequency excitation of the coils. The moving regions of converging magnetic fields serve as magnetic pistons

Deep Space Propulsion Systems 351

hurtling out blobs of plasma which are traveling very rapidly. A close analogy can be made between this type of motor and an induction motor which is uncurled to produce linear, rather than circular, motion.

Accelerators of this type have been reported to yield specific impulse values up to 12,000 seconds, but it is difficult to find any results on thrusts or power efficiency. For an engine now under design, a thrust of about one-fourth pound, with a power input of 25 kilowatts, has been predicted by A. S. Penfold, of Litton Systems, Inc. This would be a significant advance in traveling wave engines.

PHOTON PROPULSION

Photon propulsion means propulsion by formation of a beam of electromagnetic energy. Although frequent reference to this type of motor suggests that it should be commented on here, it can readily be shown to be impractical. Its impracticality is related essentially to the fact that the velocity of light lies far outside the range of exhaust velocities indicated by equation (4).

For electromagnetic radiation the ratio of energy to momentum is c (the velocity of light), independent of frequency. This is the ratio of power required to thrust which can be obtained with an engine of this type, and for this high ratio, it can readily be shown that a power supply weight of 0.01 pound per kilowatt or less would have to be achieved to yield accelerations of the order of 10^{-4} g. It is not possible to achieve this low value of α by any projection of present technology.

One can speculate on space probes to go out to our nearer stars, such as Alpha Centauri. If the trip is to be accomplished in a time of the order of ten years, a velocity of the order of the velocity of light must be given the ship. Equation (3) would then indicate that a photon exhaust is appropriate, and we can speculate as to whether there exists, even in principle, the possibility of executing such a mission if it were possible to generate the photons by the most efficient power generation, the annihilation of a portion of the vehicle mass.

The imposition of a 10-year limit to reach a velocity near

c implies an acceleration of 0.1g., requiring 10^{-5} pounds per kilowatt or less. A simple numerical calculation readily shows that to achieve this goal would require essentially complete annihilation of all matter on board the vehicle. Considering this, and the fact that no method of assembling enough anti-matter presents itself, and that if it did it would still be necessary to develop a way of controlling the annihilation rate (which appears the simplest part of all), and that the annihilation radiation is not in a form which is easily focused, it is seen that the problem of reaching even our nearest neighbor star is formidable. It may create a feeling of claustrophobia among our more ambitious astronauts but it appears that during this era of science and technology, man and his material probes are bound within the confines of his native solar system.

SOLAR SAILING

In the solar sailing technique, no shipboard power or mass for ejection is required to produce an acceleration in space. The force used is the reaction to sunlight bounced from a reflecting membrane. Before considering the question whether the accelerations obtainable are large enough to be useful, it is important to refute the common misconception that with solar sailing it is possible only to move away from the sun and never toward the sun.

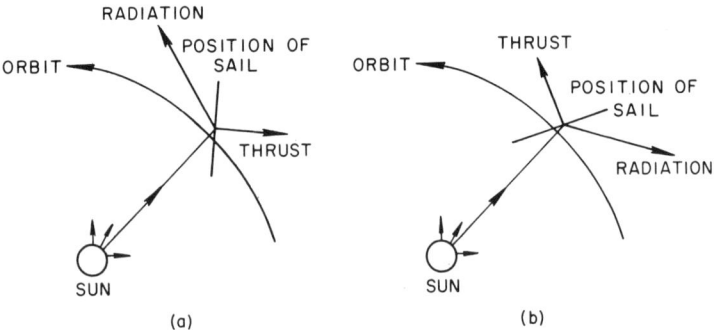

Fig. 9—(a) Solar sail orientation permitting motion toward the sun; (b) Solar sail orientation permitting motion away from the sun.

In Figure 9 two sail orientations are shown. In Figure 9a there is a component of the thrust opposing the orbital motion and the orbit of the sailing vessel will move in toward the sun. That there is also a component of thrust away from the sun is of almost no consequence. In effect one may say that the vehicle inertia has replaced the role which a keel plays for a surface sailing vessel. Figure 9b shows the vessel moving toward the sun.

To find what accelerations are obtainable, imagine that the sail is made so large that all other weights are negligible compared to the weight of the sail itself. In that case the possible acceleration is readily obtained.

$$a = \frac{\text{solar energy flux density/velocity of light}}{\text{sail mass per unit area}}$$

Taking a solar flux density as that at the earth's orbit, about 0.9×10^{-4} dynes/cm^2, and a sail made of aluminum plated mylar 0.0005 inches thick (the value used for the Echo balloon) with a mass of 0.0018 gm/cm^2, yields an acceleration of 0.05 cm/sec^2, about 5×10^{-5} g.

This acceleration is marginal and the unfurling of a very large sail with a structure of negligible weight may be very difficult. However, the lack of dependence of this method of propulsion on any large power supplies or novel types of engines which have yet to be thoroughly proven is somewhat attractive. As far as is known no serious effort to develop a vehicle based on solar sailing is under way.

CONCLUSIONS

It is clear that nuclear and electric propulsion systems offer the hope of greatly facilitating the exploration of deep space. The electrostatic ion propulsion approach, the crossed field accelerators, and the pinch accelerators are all rapidly approaching the state of development needed for space testing and mission application, with considerable promise being shown by the peristalsis accelerator. At present these various approaches are not so much competitive as they are complementary, fitting different portions of the spectrum of thrust and specific impulse which will be needed for different types of missions.

Solar sailing, as an approach to deep space missions, appears hopeful for those missions which do not go much farther from the sun than the earth's orbit. Despite this limitation, there may be a useful domain for solar sailing close to the sun, and to plants with orbits inside the earth's orbit.

AUTHORS

HUGH ODISHAW, editor of *The Challenges of Space,* is Executive Director of the Space Science Board of the National Academy of Sciences, Washington, D.C. His eleven years of academic training included work in both the humanities and the physical sciences. He received a B.S. degree from the Illinois Institute of Technology, a B.A. and M.A. degree from Northwestern University, a Sc.D. (hon.) from Carleton College, and also did graduate work at Princeton University. After teaching three years at the Illinois Institute of Technology and spending four years at the Westinghouse Electric Corporation on the company's radar activities, he served as Assistant to the Director of the National Bureau of Standards from 1946 to 1953. A member of the staff of the National Academy of Sciences since 1953, he has been Executive Director of the U.S. National Committee for the IGY, as well as of the Space Science Board, Director of the U.S. World Data Center for geophysics, and Executive Secretary of the Geophysics Research Board. His books include the *Handbook of Physics* (1958) with E. U. Condon and *Science in Space* (1961) with L. V. Berkner.

RALPH S. COOPER ("Chemical and Nuclear Rocket Propulsion" page 319) is a staff member of the Theoretical Division of Los Alamos Scientific Laboratory, Los Alamos, New Mexico. He has done research on solid state and reactor physics, electrochemistry, and nuclear rocket propulsion. In addition to his scholarly publications, he has written several satires; "New World Revisited," which appears in the April, 1962, *Bulletin of the Atomic Scientists,* outlines the perils Columbus would have faced if he had lived in an age of space and committees.

A. THEODORE FORRESTER ("Deep Space Propulsion Systems," page 336) directs work on ion propulsion as head of the ion physics department at Electro-Optical Systems, Inc., Pasadena, California. After receiving his Ph.D. in physics from Cornell University in 1942, he worked on the electromagnetic separation of uranium isotopes at the University of California and Oak Ridge. He was associate professor of physics at the University of Southern California, 1946–54. His fields of publication include isotope separation, mass spectroscopy, optics, photoelectric mixing of light, gas discharge, electrical noise phenomena, superconductivity, and ion propulsion. His organizational affiliations include Sigma Xi, Phi Kappa Phi, fellowship in the American Physical Society, and senior memberships in the Institute of Radio Engineers and the American Rocket Society.

ARNOLD W. FRUTKIN ("International Programs of NASA," page 267) is director of the Office of International Programs of the National Aeronautics and Space Administration. In 1957 he was affiliated with the National Academy of Sciences where he was Director of the Office of Information of the U.S. National Committee for the International Geophysical Year and acted as deputy for international affairs to the executive director of this committee. During 1958–59 he was also secretary to the International Relations Committee of the Space Science Board and was adviser to the Academy's delegate to the first two meetings of COSPAR.

LEO GOLDBERG ("The Sun," page 129) is Higgins professor of astronomy at Harvard College Observatory, and a member of the staff of the Smithsonian Astrophysical Observatory, where he is currently doing research in solar spectroscopy from rockets and satellites. He is vice-president of the International Astronomical Union, and a member of the Space Science Board of the National Academy of Sciences.

LEONARD JAFFE ("Communications Satellite Systems," page 44) is director of communications systems, Office of Applications, at the National Aeronautics and Space Administration. He is responsible for the NASA communications satellite research and development program.

DAVID S. JOHNSON ("Meteorological Satellites," with Harry Wexler, page 7) is deputy director of the United States Weather Bureau's meteorological satellite activities. He has been chief of the Weather Bureau's Observational Testing and Development Center. His areas of research include meteorological instrument development and observing techniques, cloud physics, and meteorological rockets and satellites.

WILLIAM W. KELLOGG ("Rocket Probes," page 108), is head of the Planetary Sciences Department of the RAND Corporation, Santa Monica, California. He is a member of the Air Force Scientific Advisory Board, and during the International Geophysical Year he was a member of the Technical Panel on the Earth Satellite Program. In 1958–59 he was chairman of the Defense Department's Advanced Research Projects Agency Meteorological Satellite Committee, and he presently serves on committees of the National Academy of Sciences for meteorological aspects of atomic radiation, high altitude rocket research, and space science. His areas of research include physics of the atmosphere, turbulence and structure of the upper atmosphere, scientific uses of rockets, satellites, and space probes.

JOSHUA LEDERBERG ("Challenges to Biology," with Aaron Novick, page 89) is professor of genetics and biology at the Stanford University School of Medicine. In 1958, he received the Nobel Prize in physiology and medicine for his research in the genetics of bacteria. The chapter in this book is based on a report prepared by Dr. Lederberg for the first International Space Science Symposium, organized by the Committee on Space Research, and held in Nice, France, in 1960.

HOMER E. NEWELL ("NASA and Space," page 178) is Director, Office of Space Sciences for the National Aeronautics and Space Administration, and has held science administrative positions for the United States Naval Research Laboratory. His research includes work on differential equations (his doctorate is in mathematics), radio propagation, and rocket exploration of the upper atmosphere.

AARON NOVICK ("Challenges to Biology," with Joshua Lederberg, page 89) is professor of biology and director of the University of Oregon's Institute of Molecular Biology. He worked on the Manhattan District Project (which developed the first atomic bomb) at the University of Chicago, and at Los Alamos. After World War II, he changed his field of interest to biology. He is an editorial board member of the *Bulletin of the Atomic Scientists*. His major research areas are: protein synthesis, biological regulatory mechanisms, and mutagenesis.

JOEL ORLEN ("Space Programs of Other Nations," page 204) is secretary of the Committee on International Relations of the Space

Science Board, National Academy of Sciences. As a foreign affairs officer for the State Department and the Atomic Energy Commission, he helped negotiate the treaty for the formation of the International Atomic Energy Agency, and has participated in intergovernmental disarmament negotiations.

JOHN R. PIERCE ("Hazards of Communications Satellites," page 60) is executive director of research communications principles division for Bell Telephone Laboratories. He is a member of the National Academy of Sciences, and a fellow of the Institute of Radio Engineers, the American Physical Society, the Acoustical Society of America, the American Astronautical Society, the American Academy of Arts and Sciences, and the British Interplanetary Society. His research has covered radio, television, electronics, acoustics, vision, mathematics, group behavior, and satellite techniques.

COLIN PITTENDRIGH ("Biology and the Space Environment," page 81) is professor of biology at Princeton University. He has served as health adviser on the problems of malaria for the Brazilian Ministry of Health and Education, and for the British Colonial Office. He is a fellow of the American Academy of Arts and Sciences, a member of the council of the American Association for the Advancement of Science, of the committee on biotrons of the National Science Foundation, of the National Academy of Sciences' committee on oceanography, and Space Science Board, and serves as adviser to the National Aeronautics and Space Administration's committee on biology. He has studied comparative and cellular physiology of daily rhythms, and evolution processes and mechanisms.

LEONARD E. SCHWARTZ ("International Space Organizations," page 241) is presently assistant to the executive director of the United States National Committee for the International Geophysical Year, acting secretary of the bioastronautics committee of the Space Science Board, and a research associate at Duke University's World Rule of Law Center. He has been a research and development program budget analyst for the United States Air Force, and director of information and education for the Army at Charleston, South Carolina.

WILLIS H. SHAPLEY ("United States Space Program," page 161)

is deputy chief of the military division in the Bureau of the Budget. Recipient of a Rockefeller public service award in 1956, he deals with defense programs, especially military research and development, and aeronautics and space programs. He is the author of "Special Problems of Military Research and Development," *Annals of the American Academy of Political and Social Science,* January, 1960.

JOHN A. SIMPSON (Introduction to Part II, "Space Research," page 75) is professor of physics at the University of Chicago's Enrico Fermi Institute for Nuclear Studies. He served as a Scientific Group Leader for the Metallurgical Laboratory at the University of Chicago (a code name for the Chicago group which worked on the atomic bomb during the second world war). He is a member of the National Academy of Sciences and of its Space Science Board, and was a member of the international Special Committee for the International Geophysical Year (CSAGI). His current research includes studies of the origin and properties of cosmic rays, magnetic fields in interplanetary space, and high-energy astrophysics.

LYMAN SPITZER, JR. ("Flying Telescopes," page 97) is Charles A. Young professor of astronomy at Princeton University, and director of the Princeton University Observatory. A theoretical astrophysicist by training, he was one of the first to become seriously interested in the possible use of satellites for astronomical observations. He is chairman of the executive committee of Princeton's Plasma Physics Laboratory, and president of the American Astronomical Society (1960–62). His publications include *Physics of Fully Ionized Gases,* and he edited a collection of papers, *Physics of Sound in the Sea.* Interstellar matter, cosmogony, stellar atmospheres, and controlled release of thermonuclear energy, are among his research interests.

GEORGE P. SUTTON ("Space Vehicles," page 304) is manager of Long Range Planning at Rocketdyne, a division of North American Aviation. He has served as chief scientist for the Advanced Research Projects Agency of the Department of Defense, and as president of the American Rocket Society. In 1958, he was J. C. Hunsaker professor of aeronautical engineering at Massachusetts Institute of Technology. He is author of *Rocket Propulsion Elements,* and co-editor of the *Advanced Propulsion Systems* (Pergamon Press, 1959).

JAMES A. VAN ALLEN ("The Earth and Near Space," page 118) is professor of physics and chairman of the physics department at the State University of Iowa, Iowa City. His satellite and rocket probe experiments revealed the existence of the now well-known Van Allen radiation belts, which surround the earth at distances of 400 and 90,000 kilometers. He is presently associate editor of the *Journal of Geophysical Research,* and a member of the Space Science Board of the National Academy of Sciences, a founding member of the International Academy of Astronautics, and a fellow of the American Rocket Society. He edited *Scientific Uses of Earth Satellites,* a collection of experiments proposed by scientists in 1956, before the first satellite had been launched.

H. C. VAN DE HULST ("COSPAR and Space Cooperation," page 291) has been president of the Committee on Space Research (COSPAR) of the International Council of Scientific Unions, from 1958 until 1962. He is professor of theoretical astronomy at Rijksuniversiteit te Leiden (State University of Leyden), Netherlands. He is also vice-president of the European Preparatory Commission for Space Research (COPERS), which plans co-operative space activities among European nations that wish to share the efforts of a scientific space program. His research work has dealt mainly with interstellar and interplanetary matter and with radio astronomy.

GERARD DE VAUCOULEURS ("The Moon and Planets," page 142) is associate professor of astronomy at the University of Texas, and a research associate at Harvard College Observatory. He is a member of the Committee on Exploration of the Moon and Planets, of the Space Science Board. He contributed a chapter to *The Exploration of Space,* a collection of papers edited by Robert Jastrow, which was published in 1960.

A. G. WAGGONER ("Department of Defense Space Program," page 195) is executive assistant to the director, Research and Systems Engineering Division, at the Airborne Instruments Laboratory. He organized the office of the assistant director, Defense Research and Engineering (Ranges and Space Ground Support), in 1960, and was made assistant director. In 1961, he became special assistant to the director, Defense Research and Engineering. A member of the American Rocket Society, he has won the William A. Jump Memorial Award and the Department of Defense Distinguished Civilian Service Award.

AUTHORS   361

HARRY WEXLER ("Meteorological Satellites," with D. S. Johnson, page 7) is director of meteorological research for the United States Weather Bureau. He has been assistant professor of meteorology at the University of Chicago and chief scientist of the United States Antarctic Program for the International Geophysical Year. He is consultant to the Space Science Board of the National Academy of Sciences, and a member of the Academy's committee to study the biological effects of radiation, the Committee for Polar Research, and the Pacific Science Board.

G. P. WOOLLARD ("Space Research and the Earth Sciences," page 24) is professor of geophysics at the University of Wisconsin, Madison, Wisconsin. He is a research associate for geophysics and oceanography at the Woods Hole Oceanographic Institute, and a fellow of the Geological Society. He was a Guggenheim fellow at Princeton, and has served in the Defense Department's Office of Scientific Research and Development as a civilian scientist. His research includes studies of the relation of gravity and magnetic anomalies to geologic structure of continents, application of geology and geophysics to engineering problems, and studies of the earth's magnetic field.

CHRISTOPHER WRIGHT ("The United Nations and Outer Space," page 277) is executive director of the Columbia University Council for Atomic Age Studies. He has done experimental physics research at Los Alamos, and helped to design the first fast nuclear reactor. He has led seminars on the social implications of science at Columbia University, and has been a research associate at the University of Chicago Law School. His recent publications include a NASA-sponsored Brookings Institution study, *Implications of Peaceful Space Activities for Human Affairs,* on which he collaborated; and "Scientific Progress as the Government of Outer Space," in the *Journal of International Affairs* (1959).

INDEX OF NAMES

Alfven, H., 119
Allison, L. J., 23
Amaldi, E., 263
Anfinsen, C. B., 96
Arrhenius, Svante, 91
Auger, Pierre, 263

Bandeen, W. R., 22
Bénard, J. A. V., 10
Berkner, L. V., 115, 355
Blagonravov, A. A., 294
Bort, Tesserenc de, 110
Bristor, C. L., 22
Buchar, E., 294
Busk, A. De, 88

Carnot, J., 311, 342
Chapman, Sidney, 119, 136
Christofilos, Nicholas, 121, 127
Clarke, Arthur C., 60, 61, 63, 71
Cocconi, G., 96
Columbus, ix, 39, 355
Condon, E. U., 358
Cooper, Ralph S., 319, 337, 355
Cowie, D. B., 96
Crick, F. H. C., 90

Demetriades, S. T., 348
Deslandres, H., 129

Edlén, Bengt, 136
Eisenhower, D., 172, 174
Eratosthenes, 28
Erickson, C., 99

Ferraro, V. C. A., 119
Forrester, A. Theodore, 339, 355
Friedman, Herbert, 138
Fritz, S., 15, 23
Frutkin, A., 267, 356

Gagarin, Yuri, 207

Galileo, 25, 129, 133
Glenn, John, xiii, 259
Goddard, Robert, 324
Goldberg, Leo, 129, 356

Hale, G. E., 129, 130
Hanel, R. A., 22
Horowitz, N., 96
Hubert, L. F., 23
Hutchings, E., Jr., 96

Ivanenko, D. D., 43
Izsak, I. G., 29, 43

Jaffe, Leonard, 44, 356
Jastrow, R., 150, 360
Johnson, D. S., 7, 284, 356
Johnson, Lyndon B., 176

Kallmann, H. K., 295
Kaufman, H. R., 347
Kellogg, W. W., 108, 357
Kennedy, John F., 71, 115, 156, 172, 174, 180, 183, 258, 259, 276, 279, 280
Khrushchev, N. S., 258, 259
Krueger, A. F., 23
Kuiper, G. P., 150

Lederberg, Joshua, 82, 89, 96, 357
Licht, J., 22
Lodge, Henry Cabot, 255

McCormack, John W., 176
McLellan, D. C., 52, 53
Mairan, De, 86
Markowitz, W., 32, 43
Massey, Sir H. S. W., 294
Middlehurst, B., 150
Miller, S. L., 96
Moore, P., 150
Morgulis, S., 96

Morrison, P., 96
Morrow, W. E., Jr., 51, 52, 53, 63

Neil, E. A., 23
Newcomb, Simon, ix
Newell, Homer E., 178, 357
Newton, Sir Isaac, 25, 27, 338
Novick, Aaron, 82, 89, 357

Odishaw, Hugh, xv, 3, 115, 155, 235, 300, 355
Oparin, A. I., 96
Ordway, F. I., III, 72
Orlen, Joel, 204, 357

Penfold, A. S., 351
Pierce, John R., 50, 60, 72, 358
Pittendrigh, Colin, 81, 358
Porter, Richard W., 293
Precht, H., 84

Rabinowitch, Eugene, viii
Roy, Maurice, 245, 294
Rush, J. H., 96
Ruzecki, M. A., 22

Sagan, C., 150
Schwartz, Leonard, 241, 358
Schwarzschild, Martin, 134
Shapley, Willis H., 161, 358
Simpson, John, 75, 359
Speiser, E., 338
Spitzer, Lyman, Jr., 97, 359

Stampfl, R. A., 22
Stanyakovich, K. P., 43
Sternberg, S., 23
Stoermer, C., 120
Stroud, W. G., 22, 23
Sutcliffe, R. C., 284
Sutton, George P., 304, 359

Titov, Gherman, 207
Tourville, L., 23

Urey, Harold C., 96, 145

Van Allen, James A., 76, 118, 359
van de Hulst, H. C., 248, 283, 291, 360
Vaucouleurs, Gerard de, 142, 150, 151, 360
Viaut, A., 22
von Karman, Theodore, 255

Waggoner, Alvin G., 195, 360
Watson, J. D., 90
Welsh, Edward C., 172
Wexler, Harry S., 7, 284, 361
Whitney, L. F., 23
Wiesner, Jerome B., 172
Winston, J. S., 15, 23
Woollard, George P., 24, 361
Wright, Christopher, 277, 361
Wright, Wilbur, ix

Zonn, W., 294

SUBJECT INDEX

Absorption lines, interstellar, 99, 105–6
Acceleration (*see also* Propulsion)
 of gravity, 25
 of particles, 123, 137
Accelerator, peristalsis, 350, 351
Advanced Research Projects Agency, 164, 165, 166, 196
Advent, 67–68, 70, 200
Aerobee, 8, 217
Aeronautics and Astronautics Coordinating Board, 170, 171, 197
Aeronomy, 110, 115, 116
Agena, 198
Agena B, 200
Airglow, 101, 102, 114, 121, 128, 216, 221, 223
Albania, 279
Albedo, of Earth, 10, 11
Albedo neutron flux, 214
Alexandria, 28
Alpha Centauri, 351
Alpha particles, 77, 78
American Telephone and Telegraph Company, 3, 58, 69–70
Amino acids, 90, 91
Andoya, 225
Angular momentum, 41
Antarctic treaty, 237
Anti-gravity, 25
Apollo project, 82, 83, 183, 334
 cost, 157
Applications, xi, xii, 3–6, 75, 179–80, 186
 Department of Defense, 200–202
 organization in United States for, 168–70
Arcas, 113, 226
Arequipa University, 225
Argentina, 210, 214, 266, 271, 272, 273, 274, 279

Argus experiment, 6, 121
Army Ballistic Missile Agency, 164
 Development Operations Division, 178
Artificial satellites, x–xiv, 75–77
 (*see also* Communications, Geodesy, etc.)
 engineering problems, 192–93, 200
 launching of, 189
 NASA program, 188–91
 schedule of U.S. launchings, 187
 as tool for planetary studies, 146, 147
 velocity requirements for orbit, 319–20
Astrolabe, 39
Astronomical observatory, orbiting, 100, 101
Astronomical Observatory of San Juan, 214
Astronomical research using space tools, 97–107
Astronomy, 97–107
 balloon investigations, 145
 and frequency allocation, 297
 high resolution research, 99–103
 observations over extended wavelength range, 103–5
 orbiting observatories, 190–91
 programs in other countries, 215, 231
 research problems, 98–103
Atlantic Missile Range, 167, 202
Atlas, 4, 8, 168, 185, 200, 202
Atlas-Able, 182, 184–85
Atlas-Agena, 166, 182, 184, 185, 187, 201, 334
Atlas-Centaur, 166, 184, 185, 306, 307, 334
Atmosphere, 7, 8, 108–15
 computer analysis of motions, 18

365

composition, 7, 109–15, 126–28, 205–6
  ground-based sources of information, 108–9
  injection of sodium and lithium, 6, 218, 222–23, 225, 228
  models of, 18, 108, 110–11, 246
  nitrogen in, 114, 115, 126
  oxygen in, 112, 113, 114, 115, 126
  planetary, 13, 144, 147–49, 192
  programs, 113, 205–6, 208, 210, 216–29, 232
  radiation emission, 18
  simulation of Earth's, 84
  of stars, 98–99, 104–5
  temperature of, 13–14, 18, 111–12, 128
  U.S. program, 7–19, 113, 186–88, 190, 191, 198, 199
Atomic Energy Commission, 161, 163, 164, 169, 170, 262
Aurora, 114, 120, 121, 123, 128, 137, 220
Australia, 271, 272, 273, 279
  space program, 214–16
Austria, 264, 279
Attitude control, 40, 190, 314

Baker-Nunn, 38, 215, 221, 224, 225, 227, 231
Ball Brothers Research Corporation, 138, 139
Balloons, 110, 140, 145, 146
Bancroft, 227
Basic research, xii, 75–80
  Moon and planets, 141, 142
  NASA program, 183
  nature of scientific research, 291
Behaviorial responses in space, 87
Belgium, 262, 263, 272, 279
  space program, 216–17
Bell Telephone Laboratories, 70
Bermuda, 231, 273
Beta 2 (1958), 38
Biology, 81–96
Black Brant, 217
Blue Streak, 264
Bratislava, 218
Brazil, 232, 272, 279
British East Africa, 272

British Meteorological Office, 227
Brno, 218
Bulgaria, 279
Bulletin of the Atomic Scientists, viii
Bureau of the Budget, 161, 170, 173–75

C-H molecules, 93
California Institute of Technology, 164
Canada, 232, 269, 270, 271, 272, 273, 274, 279
  space program, 217–18
Canadian–U.S. Meteorological Rocket Network, 116
Canton Island, 273
Cape Canaveral, 167
Cape Town, 227
Cartography, 24
Centaur, 182, 187, 305, 333
CETEX, *see* Committee on Contamination by Extraterrestrial Exploration
Chad, 257, 272
Chaff for communications, 51
Challenges, space, ix
Chemical fossils, 91
Chemical propulsion, 321–25
Chile, 231, 232, 273, 274
China, 272, 274
Chosica, 225
Circadian oscillations, 86, 87
Civilian-Military Liaison Committee, 171
Civilian space program, U.S., 181–84
Climate, 13, 14
Clouds, 8–17
  cloud cover analysis, 14–15
  data, 14–15
  photographs, 8–11
Clouds, interstellar, 97
Coaxial guns, 349, 350
Colombia, 272
Comité Spécial de l'Année Géophysique Internationale, 243
Committee on Contamination by Extraterrestrial Exploration, 250–51

SUBJECT INDEX 367

Committee on Space Research, 19, 21, 116, 170, 204, 205, 209, 210, 236, 237, 238, 243–50, 258, 269, 275, 291–98
 achievements, 295–96
 charter, 288
 evaluation, 295–96, 298
 membership, 244–46
 organization, 292–95
 role in co-operation, 242, 247–50, 275, 291–98
 and United Nations, 257, 260, 281, 287, 296–97
 working groups, 246, 251
Communications, xi, 3, 20, 21, 44–71, 201
 active repeater satellite, 56, 57
 active satellites, 45, 48, 52–55, 61, 62, 66
  orbits of, 52
  24-hour satellite, 54
 altitudes of satellites, 46, 48, 62, 67
 American Telephone & Telegraph system, 69–70
 attitude control, 54
 commercial requirements, 54, 68
 comparison of passive and active satellites, 51
 delayed repeater satellites, 56
 early theory, 60–63
 environmental space problems, 64–65
 experimental flight programs, 56–59
 frequencies, 61, 67, 68, 69
 hazards, 56, 58, 60, 65–66
 instrumentation in satellites, 58, 66–68
 international facilities, 68
 Japan's program, 224
 long-distance radio, 138
 low altitude systems, 54
 meteorological, 17, 20–21
 Moon as method of, 49–50
 NASA international program, 271
 orientation of satellites, 25
 ownership, 69–70
 passive satellites, 45, 48–52, 56, 61, 62, 66, 67, 68

 political problems, 70–71
 power, 58, 61, 65
 scattering, 63
 single satellite circuit, 54
 stationary satellite, 60–61
 synchronous satellites, 54
 24-hour satellite, 54, 60
 types of satellites for, 45–56
 U.S. organization for satellite system, 168–69
 U.S. policy, 71
 velocity control, 54
Congress
 committees of, 161
 organization of responsibilities, 175–77
 special investigating committees, 177
Contamination, 250–51
 of Earth, 95
 of planets, 94–96
Contamination, nuclear, 329
Continents, position of, 26, 29–34
Control, international, 235–36
Copernican revolution, vi
COPERS, see European Preparatory Commission for Space Research
Coriolis forces, 84
Corpuscular radiations, 118–28
Cosmic ray albedo, 127
Cosmic rays, 77–79, 80
 galactic, 78–80
 intensity, 78, 79, 206
 solar, 137
COSPAR, see Committee on Space Research
COSPAR International Reference Atmosphere, 217, 246
Costa Rica, 272
Courier, 56, 64
Crossed magnetic field engine, 349
CSAGI, see Comité Spécial de l'Année Géophysique Internationale
Curaçao, 224, 273
Czechoslovakia, 272, 279
 space program, 218

Data dissemination, 22, 250, 275–76

meteorological, 15, 16, 20, 22
  NASA policy, 193
Decontamination of spacecraft, 95
Delta, 182, 184–85
Denmark, 263, 272, 274
Deoxyribonucleic acid, 90
Department of Defense, 161, 163, 164, 165, 166, 169, 194–203
  information gathering satellites, 200–201
  inspection satellites, 200–201
  launching and range facilities, 167
  military space projects, 199–202
  NASA, relations with, 167–68, 195–97
  "piggy back" research satellites, 198
  space activities, 195–203
  space research, 197–99
  support areas, 202
  support satellites, 200–201
  upper atmosphere research program, 198, 199
  vehicle program, 202
Department of Justice, 169
Department of State, 169
Dipole belt, 51, 52, 53, 63, 66
Discoverer, x, 197–98
Discoverer XIII, 197
Discoverer XXXV, 198
DNA, *see* Deoxyribonucleic acid
Dynasoar, 166, 182, 198–99

Earth, 13–14, 24–34, 42–43, 118–22
  composition, 43
  gravitational effects, 42–43
  gravitational field, 27, 31, 33, 36, 118
  interior of, 42
  magnetic field, 43, 65, 109, 118–19
  oblateness of, 29
  pear-shaped model of, xi, 27
  7 geodetic models of, 29
  shape of, 26, 33, 42
    geodetic methods for determining, 27
  size, 33

Echo, 64, 66, 179, 219, 227
Echo I, 48, 49, 56, 218, 223, 271
Echo II, 50, 58
Ecuador, 231, 273, 274
El Salvador, 272
Electro-Optical Systems, Inc., 346
Electromagnetic energy, 351
Electromagnetic radiation, 97
Electronics, 30, 36–38
  environmental space problems, 64
  impact of technology on, 316
Electrons, 76, 77, 114, 126, 128
  (*see also* Van Allen radiation belts, particles)
  density, 114
  direction of drift, 124
  energetic, 119
Electrothermal engines, 348
Energy budget of the Earth, 13, 14
Environment (*see also* Atmosphere)
  of Earth, 85
  of Mars, 93
  terrestrial, simulation of, 84–87, 206
    atmosphere, 84
    Earth's rotation, 85–86
    gravitational field, 85–86, 87
    problems, for the engineer, 86–87
Environmental inputs, 85, 86
Equatorial-orbit satellites, 18
Eratosthenes' measurement of Earth's curvature, 28
Esselen Park, 227
Euro-Space, 262
European Launching Development Organization, 241, 262
European Preparatory Commission for Space Research, 263–64
European regional space conference, 248
European space organizations, 262–65
Europeon Space Research Organization, 241, 262
Evolution, 89
Exobiology, 82, 87, 89–96
Exploration, space, v–viii, xi, xiv, 75–76
  magnitude of U.S. effort, vii, x

SUBJECT INDEX

magnitude of U.S.S.R. effort, x
need for international co-operation, viii
Explorer, x
Explorer I, 120, 196
Explorer IV, 120, 125, 126, 127, 216
Explorer VI, 120
Explorer VII, 12, 120, 223, 224
Explorer IX, 227
Explorer X, 215
Explorer XI, 222
Extragalactic research, 98–107
Extraterrestrial life, 89–96
  planetary evaluation, 93–95
  and space probes, 93

F-1, 334
Federal Communications Commission, 169
Finland, 272
First Polar Year, 108
Fort Churchill, 108, 111, 112, 113, 119, 217
Fort Greeley, 113
France, 262, 263, 271, 272, 273, 279
  space program, 218–20
Frequency allocation, 68, 169, 281, 282, 297–98

Galactic research, 98–107
Galaxies, 97, 99, 103, 106
Gas, interstellar, 105–6
GEERS, 263
Geodesy, xi, 4, 5, 24–43
  arc measurement technique, 28
  geodetic satellites, 30–32
Geomagnetic field, 31, 76, 118–28, 137
Geomagnetic storms, 119, 122–23, 124
Geophysical observatories, orbiting, 190–91
Geophysical rockets of Soviet Union, 208–10
Germany, 232, 262, 263, 272, 273, 274
  space program, 220–21
Ghana, 257
Goddard Space Flight Center, 140, 188–89, 191

Goldstone, 215
Granulation, solar, 133–36
Gravitation, 24–31, 33–36, 40–43, 92, 102, 118
  measurement of gravitational effects, 30, 42–43
  use in guidance outside of Earth's field, 40–41
  value of g, 25
  value of G, 25
Gravitational gradient, 63, 65
Gravitational waves, 24
Gravity, international standard, 29
Gravitrons, 24
Greb, 198
Greece, 272
Ground-based observatories, 145
Guidance systems, 39–42, 312–14
  automatic inertial, 41
  and gravitational fields, 40–41
  inertial, 39, 40
  inertial celestial, 40
Gyroscope, 39, 41

Hale telescope, 102
Hardtack tests, 121
Hartebeesthoek, 227
Hawker-Siddebey, 262
Hiran, 30, 32, 37
Holmdel, N.J., 227
Honduras, 272
Hong Kong, 272
House Armed Services Committee, 176
House Committee on Science and Astronautics, 176
Huancayo Geophysical Institute, 225, 226
Hungary, 279
Hurricanes, 10, 11
Hydrogen, 106
Hydrogen, as propellant, 327–28

Iceland, 272
ICSU, *see* International Council of Scientific Unions
IGY, *see* International Geophysical Year
Independent Offices subcommittees, 177

India, 272, 273, 274, 279
  space program, 221
Industry, impact of space technology, 315–18
Inertia
  angular momentum of, 41
  in guidance, 39, 40, 313
Information exchange, COSPAR, 295
Infrared radiation, 11, 97
Infrared research, 16, 103–5
Inter-American Committee on Space Research, 265–66
Interferometer, 102
International Academy of Astronautics, 254
International Association of Geodesy, 29
International Association of Geomagnetism and Aeronomy, 249, 253
International Astronautical Federation, 248, 254–55, 260
International Astronomical Union, 249, 252, 297
International Atomic Energy Agency, 260, 261–62, 286–87
International Civil Aviation Organization, 260, 261
International co-operation, xiv, 235–40, 266, 275–76
  through COSPAR, 291–98
  in meteorology, 19–22
  need for, viii, 235, 239
  in rocketry, 115–16
  in scientific research, 236
International Council of Medical Sciences, 260
International Council of Scientific Unions, 19, 21, 204, 235, 236, 241–43, 245, 260, 280–81
  organization, 293
International Geophysical Committee, 249
International Geophysical Year, 21, 108, 110, 112, 119, 156, 170, 180, 195, 204, 205, 208, 215, 227, 237, 238, 240, 247, 267

  as precedent, xiv, 286, 287
International Institute of Space Law, 255
International Meteorological Institute, Stockholm, 226
International Meterological Organization, 19
International Meteorological Satellite Workshop, 258, 272
International programs, xiv
  (see also International co-operation)
  NASA, 267–76
International Rocket Intervals, 249, 296
International Rocket Weeks, 116
International Scientific Radio Union, 249, 252, 253, 281, 283, 297
International scientific unions, 251–54, 291
International space organizations, 241–66
International Telecommunications Union, 69, 71, 236, 241, 260, 281–83, 297
  CCIR, 281–82
International Union of Biochemistry, 252
International Union of Biological Sciences, 252
International Union of Geodesy and Geophysics, 249, 252, 253
  specialized associations, 253
International Union of Physiological Sciences, 252
International Union of Pure and Applied Chemistry, 252
International Union of Pure and Applied Physics, 249, 252, 253, 254
International Union of Theoretical and Applied Mechanics, 252, 253
International University Program, NASA, 274
International Year of the Quiet Sun, 237, 249

SUBJECT INDEX 371

Interplanetary space, density of, 118
Interstellar gas research, 105–6
Interstellar matter, 97, 99, 102
Ionization, 113, 115
Ionosphere, 44, 111, 112, 114–16, 138
   programs in other countries, 210, 216, 217, 220, 225, 226, 228, 229
   U.S. program, 186–88, 192, 199
IQSY, *see* International Year of the Quiet Sun
Iran, 231, 273, 279
Ireland, 272
Israel, 272, 274
   space program, 222
Italy, 262, 263, 270, 271, 272, 273, 274, 279
   space program, 222

Japan, 271, 272, 273, 274, 279
   space program, 223–24
JATO units, 322
Jet Propulsion Laboratory, 164, 215, 227
Jodrell Bank, 227, 229, 230, 273
Johannesburg, 215
Joint Committee on Atomic Energy, 176
Juno II, 182, 184–85
Jupiter, planet, 145, 187, 338
Jupiter C, 182, 184–85

Kappa rockets, 223
Kennedy-Khrushchev exchange, 258–59
Kiruna, 226
Kitt Peak National Observatory, 170
Kiwi, 309, 334–35
Kracow, 226

Laika, 206
Latin America, space organization in, 265–66
Lebanon, 279
Lewis Laboratories, 347
Life (*see also* Biology)
   definition of, 89,
   extraterrestrial, 82, 89, 92–93, 95
   minimal conditions for, 92
   origin of, 91
   support systems, 89, 90, 92, 206, 207
Light pressure, 63
Light-reflecting layer, 6
Lincoln Laboratories, 51, 52
Litton Systems, Inc., 351
Loki II, 113
Loran, 36, 37
Lunar exploration, *see* Moon
Lunik I, 120, 211
Lunik II, 205, 207, 212, 218
Lunik III, x, 120, 206, 212

Magnetic axis, 121, 122
Magnetic fields, 80, 84, 147
   (*see also* Geomagnetic field)
   interstellar, 97
   of Moon, 207
Magnetic screening, 78
Magnetohydrodynamic thrustors, 348–51
Magneto-hydrodynamics, 119
Man in space, xi, xii, 76, 82–88
   NASA program, 183
   scientific aspect, 159–60
   Soviet program, 206–7
Mars, xii, 41, 82, 83, 102, 143, 147, 148, 149, 187, 192, 251, 338
   environment of, 93
   life on, 93, 94
   propulsion required for mission, 331
Maser, 66, 67
Mauritius, 272
Mercury, planet, 94, 148, 193
Mercury program, x, 82, 166, 183, 202
Mesosphere, 111, 112
Meteorological rocket network, 113
Meteorological satellite system, 16–19
Meteorology, xi, 4, 7–23, 109–13
   balloons, use of, 18, 110
   India's program, 222
   international co-operation in, 19–22, 116–17, 232

NASA international program, 271–72
South Africa's program, 227
and United Nations, 283–84
U.S. organization for satellite system, 168
weather prediction, 7–19, 116–17
Meudon, 219
Mexico, 231, 232, 273, 279
Micro-environments, 85
Microlock, 38
Micrometeorites, 230
Midas, 157, 200
Military space activities, U.S., *see* Department of Defense
Minitrack, 37, 215, 218, 225, 227
Modalities, sensory, 85
Molecules, C-H, 93
Mongolia, 257
Moon, xii, 142–51, 251
  application of in communications, 49, 50
  exploration of, 142–43, 150, 159, 207
  far side of, 206
  geodetic interest, 32, 33
  and guidance system, 314–15
  life on, 94
  NASA program, 191
  and navigation, 35
  observation techniques, 143–45
  observing stations for study of, 145–46
  orbital period, 25
  propulsion required for mission, 331, 332
  radar study of, 145
  soft landings, 148–49
  velocity requirements for observation, 320
Moon camera, 32–33
Moonwatch, 214, 216, 227, 231
Mount Palomar, 102
Mount Wilson and Palomar Observatories, 134
Mount Wilson solar observatory, 129
Mutation rates, 88
Mutations, 89, 90

NASA, *see* National Aeronautics and Space Administration
National Academy of Sciences, 31, 170, 195, 247, 275
National Advisory Committee for Aeronautics, 164, 178, 181
National Aeronautics and Space Act of 1958, 165, 170, 174, 176, 178, 196, 267
National Aeronautics and Space Administration, x, xi, 10, 70, 102, 138, 140, 161, 164, 178–94, 198, 199, 215, 216, 217, 218, 222, 223, 247–48
  aeronautics, 181–83
  basic research program, 183
  budget summarized, 157–58
  and Department of Defense, 167–68, 195–97
  experiments, engineering of, 189–90
  international programs, 267–76
    exchange of information, 275–76
    ground based support, 271–72
    guidelines, 267–68
    overseas facilities, 273
    rockets and satellites, 269–71
    technical training, 273–75
  Moon and planets, 191–93
  observatory satellites, 187, 189
  organization, 180–81
  philosophy, 178–79
  publication of results, 193
  research opportunities, 194
  research satellites, 188–91
  responsibilities, 165–70
  sounding rocket program, 182
  space program
    Moon and planets, 191–93
    research satellites, 188–91
    space sciences program, 186–88
    supporting research and technology, 183–86
  space vehicles, 182–85, 334
National Aeronautics and Space Council, 161, 170, 171, 172
National Bureau of Standards, 226
National Meteorological Center, 17

SUBJECT INDEX

National Physical Laboratory, New Delhi, 222
National programs, xiv, 154–56, 204–32 (*see also* under name of country)
National Science Foundation, 164, 170
Naval Research Laboratory, 138, 140, 178
Navigation, 4, 5, 34–39, 201
  bicoordinate, 84
  by birds, 84
  celestial, 35, 39–42
  and gravity, 35
  U.S. organization for satellite systems, 168
Near space environment, 118
Netherlands, 262, 263, 272
  space program, 224
Netherlands Antilles, 272
Netherlands New Guinea, 272
Neutrons (*see also* Van Allen radiation belts)
  albedo flux, 214
  albedo hypothesis, 127
  from cosmic rays, 126
New Zealand, 156, 272, 274
Nigeria, 272, 273
Nike-Cajun, 217, 222, 225
Nimbus, 4, 16–19, 168
Northrop Corporation, 348
Norway, 263, 271, 274
  space program, 225
Nova, 166, 182
Novae, 99, 105
Nuclear energy, v
  auxiliary power supply, 332
  in deep space probes, 341
  as power source, 310, 311
  as propellant, 310, 326–41
Nucleic acids, 90, 91, 93
Nucleotides, 90
Numerical methods, meterology, 18
Nyasaland, 272

Office of Civil Defense and Mobilization, 169
Office of Emergency Planning, 169
Office of Science and Technology, 170, 172

Office of Scientific Research and Development, 162
Olifantsfontein, 227
Orbits, 25–34, 42–43
  analysis of, 26–27, 30
  determination of, 38
  studies, 218, 220, 230
  studies of perturbations of, 30
Ottawa, 218

Pacific Missile Range, 167, 202
Pakistan, 271, 274
Panspermia, 91, 94
Particles (*see also* Van Allen radiation belts)
  auroral, 114
  charged, 76, 77, 114, 118–28
    intensity, 127–28
    local acceleration, 123
  from exploding stars, 97
  trapped, 77, 118–28
Peaceful Uses of Outer Space, Committee on, 256–58, 259, 277–80
Pennsylvania State University, 226
Periodicities, 86–87
Perth, 215
Perturbations of satellites, 26, 27
Peru, 273
  space program, 225–26
Pezinck, 218
Philippines, 232
Photon propulsion, 351–52
Photons, 105
Physical Research Laboratory, Ahmedabad, 222
Physiology
  ecological, 83, 85
  environmental, 83, 84–6
Pioneer, x
Pioneer I, 120
Pioneer III, 120, 125, 126, 127
Pioneer IV, 120
Pioneer V, 79, 140, 227
Planetology, 148
Planets, xii, 102, 142–51
  atmospheres of, 144, 192
  carbon compounds, 94
  Earth-based studies, 149–50
  exploration of, 149–50, 159
  infection of, 94–96, 251

NASA program, 191–93
navigation, 35
observation techniques, 143–45
observing stations for study of, 145–46
organic matter on, 94
photoelectric observations of, 144
polarimetric observations of, 144
potential experiment using space tools, 147–49
radar study of, 145
radio observations of, 144
radiometric observations of, 144
soft landings, 148–49
spectroscopic observations of, 144
velocity requirements for probes, 320
Plasmas, 80, 109
Pluto, 338
Poland, 272, 279
  space program, 226
Polar-orbit satellites, 18, 47
Portugal, 272
Position fixing, 39
Power sources, 310–12
Power supplies (see also Propulsion)
  NASA program, 183
Praha, 218
President's Science Advisory Committee, 172–73
Pretoria, 227
Prince Albert, 218
Princeton University, 191
Princeton University Observatory, 102
Project West Ford, 51, 52, 58, 70, 169, 200, 202
Propulsion, 308–10, 319–354
  accelerating voltages, 344
  acceleration, 353
  acceleration required, 337–41
  chemical, 321–25, 332, 334, 335, 338, 339, 340
    specialized systems, 326
    and nuclear, 319–54
    comparison of methods, 310, 331–32
  deep space systems, 336–54

electric propulsion, 340, 341–42
electric systems, 309–10, 332
electrothermal engines, 348
exhaust velocity, 323, 325, 326, 328, 330, 331, 336, 338, 340, 344
gas discharge ion engine, 347
ion motors, 343–48
liquid propellants, 308, 323–25, 327, 332, 333
magnetohydrodynamic thrustors, 348–51
NASA program, 183
nuclear, 326–30, 332, 341
  advanced nuclear propulsion, 330
  electrodes, effect on, 346
  hydrogen, 327–28
  ion mass, 344–46
  neutralization problem, 345–46
  source of ions, 346
nuclear reactor, 328
nuclear rockets, 308–9
photon propulsion, 337, 351–52
power sources, 341–43
principles, 320–21
solar radiation, 341
solar sailing, 336, 352–54
solid propellants, 321–23, 327, 334, 335
summary of U.S. program, 334
thermonuclear, 330
velocity requirements, 319–20
Protein-nucleic acid system, 92
Proteins, 90, 93
Protons, 76, 77, 78, 114, 119, 124, 126, 128 (see also Van Allen radiation belts)

Radar, 37, 145
Radar observations of clouds, 8
Radiation, as hazard, 56, 65, 77–78 (see also under specific subjects)
Radiation belts, see Van Allen radiation belts
Radiation sensors, 12
Radio Research Laboratories, Kokobunji, 224

SUBJECT INDEX

Radio Research Station, Slough, 227
Radiosonde, 110
Rand Corporation, 108
Ranger, x
Rebound, 58
Receivers, for communications satellites, 67
Reconnaissance, 5
Redstone, 182, 184–85, 196, 324
Regulation, international, 235–36, 297
Relay, 58, 70
Relay systems, microwave, 44
Republic Aviation, 348
Research, xi-xiii, 75–80, 108–10, 141–42, 183, 236, 291
Resolute, 218
Revolution, scientific, v–viii
Rhodesia, 272
Ring currents, 124
Rockets, xiii, 108–17 (*see also* names of rockets)
   international co-operation, 116–17, 270–71
   NASA program, 186–88
   program in other countries, 208–19, 222–32, 269–71
   propulsion principles, 320–21
   synoptic, 113
Rockoons, 114
Rover, 169
Royal Observatory, Edinburgh, 227
Royal Society, 215
Rumania, 279

S-16 and S-17 astronomical satellites, 138–39
Saint rendezvous, 200
Samos, 157, 200
Sardinia, 222, 223
Saturn, 166, 182, 184–85, 187, 192, 305
Saturn C-1, 308, 334
Saturn C-5, 334
Science in Space, 115
Scientific research, tradition of co-operation, 236 (*see also* Research)
Score, 56, 64

Scout, 166, 182, 184–85, 188, 323, 334, 335
Second International Space Science Symposium, 204
Senate Armed Services Committee, 176
Senate Committee on Aeronautical and Space Sciences, 176, 195
Senate Subcommittee on Reorganization and International Organizations, 162
SEREB, 262
Sextant, 35, 39
Shavit II, 222
Shoran, 32, 36
Sierra Leona, 257
Skalnate Pleso, 218
Skylark, 215, 216, 228–31
Smithsonian Astrophysical Observatory, 191, 221, 227
SNAP, 169, 201, 342
Sodium injection in upper atmosphere, 6, 218, 219, 222, 223, 228
Sofar, 33, 34
Soft landings, 93
   on Moon, 148–49
   on planets, 148–49
Solar, *see* Sun
Solar Radiation III, 224
Sounding rockets, *see* Rockets
South African Weather Bureau, 227
Soviet Union, xiii, xiv, 277–79, 288
   rocket instrumentation, 208–10
   space program, 205–10
Space organizations
   international, 241–66, 291–98
   regional, 241, 262–66, 276
   specialized agencies, 258–62
Space probes, 83, 146–48
   limitations, 352–53
   propulsion, 336–54
   schedule of U.S. launchings, 187
   of Soviet Union, 207–8
   as tool for planetary exploration, 146–48
Space Research, 247, 295
Space Science Board, 31, 170, 248, 275

statement on man in space, 159–60
Space science symposia, 246–47
Space vehicles, 304–18 (*see also* Technology)
  recovery, 314
Spacewarn, 228, 248, 275
Spain, 231, 263, 273
Spontaneous generation, 91
Spores, migration of, 91, 92
Sputnik I, x, xiii, 205, 211, 216
Sputnik II, 205, 206, 211, 216, 308
Sputnik III, 120, 205, 211, 220, 224, 229, 308
Sputnik IV, 206, 212, 218, 223, 224, 308
Sputnik V, 206, 212, 218, 224
Sputnik VI, 206, 212, 224
Sputnik VII, xiii, 206, 207, 212
Sputnik VIII, 207, 212
Sputnik IX, 206, 207, 212
Sputnik X, 206, 207, 212
Stars, 97–107
  abundance of elements, 99
  atmospheres of, 98–99, 104–5
  brightness of, 104
  constitution of, 98
  distances, 103
  evolution of, 98
  geodetic interest, 33
  high resolution, need for, 102
  mass, 103
  navigation, 35
  particles from exploding stars, 97
  structure of, 104
  subsystems, 103
  systems, 99, 103
  variable, 105
State University of Iowa, 114
Stationary satellites, *see* Communications
Stratoscope, project, 135
Stratoscope II, 102
Stratosphere, 111, 112
Stratospheric observatories, 146
Sudan, 272
Sun, 97, 102, 129–35
  balloon studies of, 140
  chromosphere, 136–37
  corona, 105, 136–37
  flares, 78, 122, 137, 138
  gases, 123, 130, 136
  general characteristics of, 131–33
  granulation, 133–36
  ground-based observations, 131, 140
  instruments for study of, 129–31
  interior of, 131, 136
  navigation, use in, 35
  photosphere, 133, 134
  plasma, 80, 128
  as power source
    batteries, 31, 56, 311, 332–33
    in propulsion, 341
  rocket studies of, 146
  solar observatories, orbiting, 138, 190–91
    instrumentation, 140
  solar physics, 78–80, 129–35
  solar radiation, 7, 10–13, 94, 109, 110, 113, 114–16, 139
  solar sailing, 336, 352–54
  solar system, origin of, 143
  solar wind, 79, 80, 123
  spectrum, 115, 131
  temperature of, 133, 134, 136
Sunspot Minimum Ionospheric Rocket Sounding Program, 249
Sunspots, 132–34
  11-year cycle, 137
Surveying, 4, 5
Super-Schmidt Camera, 214
Supernovae, 99, 105
Sweden, 262, 263, 271, 274, 279
  space program, 226
Switzerland, 263, 272
Syene, 28
Symposium on Space Research, Buenos Aires, 214, 265
Synchronous satellite, 45
Synchrotron radiation, 106
Syncom, 58, 169

Taiwan, 232
Talera, 225
Technical University of Denmark, 225
Technology, xiii, 75–76, 300–303 (*see also* Propulsion, NASA, etc.)

SUBJECT INDEX

auxiliary power supplies, 332
ballistic missiles, 305, 312
 guidance, 39–41, 312–14
  attitude control, 314
  inertial system, 313
  radio guidance, 313
  sensors, 313
 impact on industry, 315–18
  electronics, 316
 impact on life sciences, 82
 influence on United States economy, 318
 liquid propellant rocket, 323
 materials, 307, 314–16, 323, 328
 power sources and conversion, 310–12
 propulsion, 308–10
 propulsion systems, 301–3, 325
 recovery of vehicles, 314–15
 scientific significance, 75–82
 significance to astronomy, 97–107
 of space vehicles, 195–97, 202, 304–7
  ground support, 307–8
  materials, 307
  materials for nozzles, 323
  payload capacity, 305–7
  structures, 307–8
Telephone and telegraph service, 69
Telephony, 3
 transoceanic, 63
Telescopes
 balloon-borne, 103–5
 in space, 98, 99–103, 106–7
Television, 60, 71
 transoceanic, 64
Temperature of the earth, 13, 14
Thailand, 272
Thermal noise, 67
Thermonuclear energy, 330
Thor, 168, 198, 202
Thor-Able, 182, 184–85
 payload capacity, 308
Thor-Agena, 166, 182, 187
 payload capacity, 308
Thor-Agena B, 184–5, 187, 334
Thor-Delta, 166, 187
Thule, 113, 116

Tiros, x, xi, 10, 14, 15, 16, 168, 227, 232
Tiros I, 4, 8, 10, 218, 222
Tiros II, 4, 12, 15, 271
Tiros III, 4, 9, 12
Tokyo Astronomical Observatory, 224
Topside sounder, 217, 218
Torun, 226
Tracking, 31–39, 223, 231–32
 coordinates, 35
 and Doppler shift, 38
 electronic methods, 36–38
 geodetic satellites, 34
 optical, 36, 38, 216, 218, 219, 221, 222, 224, 227
 radio, 215, 218, 222
 technology, 31–32, 315–16
 tracking stations, 31–32, 214, 225–28, 231, 273
Transcontinental communications, 62
Transit, 168, 198, 200, 201
Transit II-A, 138
Transoceanic communications, 62
Trinidad, 273
Turkey, 274

Ultraviolet radiation, 97, 103–5, 106, 109, 112, 114, 115
UNESCO, see United Nations Economic, Social and Cultural Organization
Union of South Africa, 215, 272, 273
 space program, 227
United Arab Republic, 272, 279
United Kingdom, 232, 262, 263, 269, 272, 273, 274, 279
 space program, 227–31
United Kingdom Ministry of Aviation, 215
United Nations, xiv, 71, 235, 236, 255–58, 260, 277–90
 Committee on Peaceful Uses of Outer Space, 277–80
 communications, 281–83
 and COSPAR, 296–97
 international science and outer space, 280–81
 meterological satellites, 283–84

outlook for coordination, 285–87
politics and problems, 228–90
specialized agencies, 280–81
United Nations Economic, Social and Cultural Organization, 19, 241, 260, 261, 280–81
United Nations' Registry, 258
United States Air Force, 157, 165, 198
United States Army, 157, 165
United States
  applications of space program, 168–70, 179, 200–202
    communications, 168–69
    meteorology, 168
    navigation, 168
  civilian space effort, 181–94
  communications satellite policy, 71
  congressional organization, 175–77
  co-ordination of, 170–75
  Department of Defense space program, 195–205
  division of responsibilities of, 164–70
  impact of space technology on economy, 318
  men in space program, 158–60
  military aspects of space program, 163
  NASA, 178–94
  NASA space science program, 186–88
  objectives, 178
  organization of space program, 161–77
  research satellites (NASA program), 188–91
  schedule of launchings, 187
  space efforts, summarized, 156–57
United States Meteorological Network, 113
United States Navy, 21, 138, 157, 165
United States Weather Bureau, 110, 161, 164, 168, 199, 248
Universe, extent of, 103
University of Wisconsin, 191
Upper Atmosphere, *see* Atmosphere

Uttar Pradesh State Observatory, 221

V-2 rocket, 8
Van Allen radiation belts, xi, 56, 57, 76–77, 88, 114, 118–28
  inner zone, 121–25, 127, 128
  inner zone composition, 126
  outer zone, 121–25, 128
  neutron decay and inner zone, 126–27
  satellite findings, 120–22
  time fluctuations in intensity, 124
  trapped radiation, origin of, 125–27
  two zones, 125–27
Vanguard, 164, 170, 178, 179, 182, 184–85, 195
Vanguard I, xi, 5, 215, 216, 229
Vanguard II, xi, 4
Vehicles, 304–18, 326
Vela-Hotel program, 165, 200, 201
Venus, xii, 1, 144, 147, 148, 187, 192, 207, 208, 338
  life on, 94
Veronique, 218, 219, 264
Viet Nam, 272
Villa Delores, 214
Vostok I, 205, 206, 207, 212
Vostok II, 206, 207, 212

Wallops Island, 223
Weapons systems, 5
Weather, *see* Meteorology
Weather control, 6
West Ford, *see* Project West Ford
West Indies, 272
White Sands Missile Range, 202
White Sands Proving Ground, 108, 110
Winds, balloon determination of, 18
WMO, *see* World Meteorological Organization
Woomera, 214, 215, 231
World Data Center A, 22
World Data Centers, 22, 250, 275
World Health Organization, 260, 261
World Magnetic Service, 249

SUBJECT INDEX

World Meteorological Organization
 19, 21, 22, 110, 236, 241,
 258, 260, 283–84, 296

X-15, 166, 182, 198-99

X-ray research, 103–5
X-rays, 97, 109, 138

Zanzibar, 273

DATE DUE			
MAY 13			
MAY 10 1965			
OCT 27			
MAR 3			